Mechanisms of Yeast Recombination

Edited by

Amar Klar
Cold Spring Harbor Laboratory

Jeffrey N. Strathern
NCI-Frederick Cancer Research Facility

Titles in
Current Communications in Molecular Biology
PLANT INFECTIOUS AGENTS
ENHANCERS AND EUKARYOTIC GENE EXPRESSION
PROTEIN TRANSPORT AND SECRETION
IMMUNE RECOGNITION OF PROTEIN ANTIGENS
EUKARYOTIC TRANSCRIPTION
PLANT CELL/CELL INTERACTIONS
TRANSLATIONAL CONTROL
COMPUTER GRAPHICS AND MOLECULAR MODELING
MICROBIAL ENERGY TRANSDUCTION
MECHANISMS OF YEAST RECOMBINATION

MECHANISMS OF YEAST RECOMBINATION
©1986 by Cold Spring Harbor Laboratory
All rights reserved
International Standard Book Number 0-87969-195-6
Book design by Emily Harste
Printed in the United States of America

The individual summaries contained herein should not be treated as publications or listed in bibliographies. Information contained herein can be cited as personal communication contingent upon obtaining the consent of the author. The collected work may, however, be cited as a general source of information on this topic.

All Cold Spring Harbor Laboratory publications may be ordered directly from Cold Spring Harbor Laboratory, Box 100, Cold Spring Harbor, New York 11724. (Phone: Continental U.S. except New York State 1-800-843-4388. All other locations [516] 367-8425.)

Conference Participants

James R. Broach, Dept. of Molecular Biology, Princeton University, New Jersey
Ronald A. Butow, Dept. of Biochemistry, University of Texas Health Science Center, Dallas
Michael M. Cox, Dept. of Biochemistry, University of Wisconsin, Madison
Richard Egel, Institute of Genetics, University of Copenhagen, Denmark
Michael Esposito, Donner Laboratory, University of California, Berkeley
Rochelle Easton Esposito, Dept. of Molecular Genetics and Cell Biology, University of Chicago, Illinois
Seymour Fogel, Dept. of Genetics, University of California, Berkeley
David J. Garfinkel, NCI-Frederick Cancer Research Facility, Maryland
Herbert Gutz, Institut für Genetik, Technische Universität, Braunschweig, Federal Republic of Germany
James Haber, Rosenstiel Center, Brandeis University, Waltham, Massachusetts
P.J. Hastings, Dept. of Genetics, University of Alberta, Edmonton, Canada
Fred Heffron, Dept. of Molecular Biology, Scripps Clinic and Research Foundation, La Jolla, California
Makkuni Jayaram, Dept. of Molecular Biology, Scripps Clinic and Research Foundation, La Jolla, California
David Kaback, Dept. of Microbiology, UMDNJ-New Jersey Medical School, Newark
Amar Klar, Cold Spring Harbor Laboratory, New York
Hannah Klein, Dept. of Biochemistry, New York University Medical Center, New York
Richard Kolodner, Dana Farber Cancer Institute, Boston, Massachusetts
Sam Kunes, Dept. of Biology, Massachusetts Institute of Technology, Cambridge
Robert Malone, Dept. of Biology, University of Iowa, Iowa City
Robert Mortimer, Dept. of Biophysics, University of California, Berkeley
Alain Nicolas, Massachusetts General Hospital, Boston
Thomas D. Petes, Dept. of Molecular Genetics and Cell Biology, University of Chicago, Illinois
Michael A. Resnick, National Institute of Environmental Health Sciences, Research Triangle Park, North Carolina
Shirleen Roeder, Dept. of Biology, Yale University, New Haven, Connecticut
Herschel L. Roman, Dept. of Genetics, University of Washington, Seattle
Paul D. Sadowski, Dept. of Medical Genetics, University of Toronto Faculty of Medicine, Canada

Brian Sauer, E.I. du Pont de Nemours & Company, Wilmington, Delaware
Franklin W. Stahl, Institute of Molecular Biology, University of Oregon, Eugene
Jeffrey N. Strathern, NCI-Frederick Cancer Research Facility, Maryland
John Wallace, Dept. of Human Genetics and Development, Columbia University College of Physicians & Surgeons, New York

Preface

This meeting on Mechanisms of Recombination in Yeast at the Banbury Center of Cold Spring Harbor Laboratory brought together some 30 experts in various aspects of recombination and gene conversion. They included classical geneticists, molecular geneticists, molecular biologists, and biochemists with interests ranging from the enzymology of site-specific recombinases to the genetic consequences of general recombination. Their discussions addressed both meiotic and mitotic events. As a result, the meeting was a terrific experience. The talks were all excellent and merited the close attention and critical thinking that this group was able to provide. The informal environment and the intimate size of the meeting encouraged spirited discussion. And, of course, the spirits after dinner encouraged informal discussions.

We are grateful to our colleagues who advised us regarding the content of the meeting and those who came and joined in the rather lively proceedings. We thank Jim Watson for his support and continued interest in the topic; Mike Shodell, Director of Banbury, and Beatrice Toliver for organizing the details of the meeting; Katya Davey for creating at Banbury an atmosphere that stimulates spirited late night discussions; and Judy Cuddihy and Nancy Ford, Director of Publications, for their help in producing this book soon after the meeting.

A.J.S.K.
J.N.S.

The meeting on Mechanisms of Yeast Recombination was funded entirely by proceeds from the Laboratory's Corporate Sponsor Program, whose members provide core support for Cold Spring Harbor and Banbury meetings:

Abbott Laboratories
American Cyanamid Company
Amersham International plc
Becton Dickinson and Company
Cetus Corporation
Ciba-Geigy Corporation
CPC International Inc.
E.I. du Pont de Nemours & Company
Eli Lilly and Company
Genentech, Inc.
Genetics Institute
Hoffmann-La Roche Inc.
Monsanto Company
Pall Corporation
Pfizer Inc.
Schering-Plough Corporation
Smith Kline & French Laboratories
The Upjohn Company
Wyeth Laboratories

Contents

Conference Participants, iii
Preface, v

Interaction of FLP Protein with its Recombination Site   1
R.C. Bruckner, J.F. Senecoff, L. Meyer-Leon, and M.M. Cox

Interaction of the FLP Recombinase of the 2-Micron Plasmid with its Target Sequence   7
P.D. Sadowski, B.J. Andrews, L.G. Beatty, D. Sidenberg, and G. Proteau

Site-specific Recombination Promotes Plasmid Amplification in Yeast   11
F.C. Volkert and J.R. Broach

Double-strand-break Repair in Yeast Results in Conversion Events That Resemble Mating-type Switching   19
M. Jayaram

Recombination of Mitochondrial Genes   25
R.A. Butow, P.S. Perlman, and A.R. Zinn

A Recombination-stimulating Sequence in the Ribosomal RNA Gene Cluster of Yeast   29
G.S. Roeder, R.L. Keil, and K.A. Voelkel-Meiman

Switching Genes in *Schizosaccharomyces pombe* and DNA Rearrangements in the Mating-type Region of *S. pombe*   35
H. Gutz, L. Heim, P. Kapitza, and H. Schmidt

Relationship of DNA Breakage at the *smt* Site and Mating-type Switching in *Schizosaccharomyces pombe*   41
R. Egel

Initiation and Resolution Steps of Recombination for Yeast Mating-type Interconversion   47
A.J.S. Klar

Intermediates in Homothallic Switching   53
J. Strathern, C. McGill, B. Shafer, and D. Raveh

Stimulation of Mitotic Recombination by *HO* Nuclease   61
J.A. Nickoloff, E. Chen, and F. Heffron

Double-chain Breaks: Thinking about Them in Phage and Fungi   69
F.W. Stahl, D.S. Thaler, A. Kolodkin, S. Rosenberg, and E. Sampson

The Role of *RAD50* in Meiotic Intrachromosomal Recombination   75
J. Wagstaff, S. Gottlieb, and R. Easton Esposito

Screening for Recombination-defective Mutants with a Positive Selection System for Plasmid Excision   85
R.H. Schiestl and P.J. Hastings

Genetic and Molecular Analyses of Recombination Using Mutants Altered in DNA Repair and Sister Chromatid Recombination   89
M.A. Resnick, A.M. Chaudhury, and J.L. Nitiss

A Site for the Initiation of Gene Conversion in Meiosis   95
R.E. Malone, S. Cramton, and R. Gehrhardt

REC Genes Governing Mitotic Recombination, Chromosomal Stability, and Sporulation: Cell Type and Life Cycle Stage-specific Expression of *rec* Mutants   103
M.S. Esposito, K. Bjornstad, L.L. Holbrook, and D.T. Maleas

Association of Reciprocal Exchange and Intrachromosomal Gene Conversion in Mitosis   111
H.L. Klein and K.K. Willis

High-frequency Meiotic Recombination Events Do
Not Require End-to-end Chromosome Synapsis 117
S. Jinks-Robertson and T.D. Petes

Meiotic Recombination between Dispersed
Homologous Sequences in *Saccharomyces cerevisiae* 123
M. Lichten, R.H. Borts, and J.E. Haber

Genetic Control of Delta Recombination 131
J.W. Wallis, G. Chrebet, A. Beniaminovitz, and R. Rothstein

Detecting Heteroduplex DNA in Postmeiotic
Segregation: And Recombination in a Nontandem
ADE8 Duplication 137
S. Fogel, J.W. White, A. Plessis, and D. Maloney

Gene Conversion and Associated Recombination in
Heterozygous Versus Heteroallelic Diploid Cells 141
H. Roman

Synapsis-dependent Illegitimate Recombination and
Rearrangement in Yeast 149
S. Kunes, D. Botstein, and M.S. Fox

Is There Distributive Pairing in *Saccharomyces
cerevisiae*? 157
D.B. Kaback

Enzymatic Systems from *Saccharomyces cerevisiae*
That Catalyze the Processing of Holliday Junctions
and the Repair of Mismatched Nucleotides 165
R. Kolodner, D. Evans, L.S. Symington, and C. Muster-Nassal

Ty Element Retrotransposition 173
D.J. Garfinkel, J.D. Boeke, and G.R. Fink

Gene Conversion Mechanisms of Punctual and Nonpunctual Mutations in *Ascobolus* 181
A. Nicolas, H. Hamza, and J.-L. Rossignol

SUMMARY
J.N. Strathern 189

Interaction of FLP Protein with Its Recombination Site

R.C. Bruckner, J.F. Senecoff, L. Meyer-Leon, and M.M. Cox
Department of Biochemistry, College of Agriculture and Life Sciences
University of Wisconsin-Madison, Madison, Wisconsin 53706

The FLP protein promotes a site-specific recombination event that results in the inversion of specific sequences within the yeast 2-micron plasmid (Broach et al. 1982). The protein is encoded by the 2-micron plasmid and has a predicted molecular weight of about 48,000 (Hartley and Donelson 1980), and it has been purified extensively (Babineau et al. 1985; L. Meyer-Leon et al., unpubl.). This recombination event has been characterized both in vivo (Broach et al. 1982; Jayaram 1985) and in vitro (Meyer-Leon et al. 1984; Sadowski et al. 1984). Much of the work to date has focused on the recombination site.

The FLP Recombination Site

Our current understanding of the sequences that define the minimal site required for recombination in vitro, and the functions of subsets of that site, is presented in Figure 1. The site illustrated consists of two 13-bp repeats inverted about an 8-bp spacer. Two base pairs may be deleted from either end without affecting any identified FLP protein contact points (R.C. Bruckner and M.M. Cox,

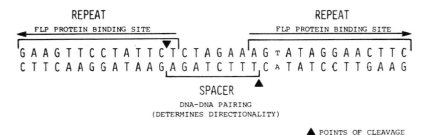

Figure 1 The FLP recombination site. See text for details.

unpubl.). Three base pairs may be deleted from either end without affecting the activity of the site in vitro (Senecoff et al. 1985). The minimal site therefore consists of the central 28 bp of the 34-bp sequence illustrated. The boundaries are not strictly defined, with reduced reactions occurring with smaller sites (Senecoff et al. 1986). The minimal site that will support detectable recombination may be as small as 20 bp (Gronostajski and Sadowski 1985a). The FLP protein cleaves this site at the boundaries of the spacer as shown, becoming covalently linked to the DNA via a 3′-phosphotyrosyl linkage, and leaving 8-nucleotide protruding ends with 5′-hydroxyl termini (Andrews et al. 1985; Gronostajski and Sadowski 1985b; Senecoff et al. 1986; R.C. Bruckner and M.M. Cox, unpubl.).

This entire site is protected from DNase digestion by FLP protein (Andrews et al. 1985). A simple functional distinction can be made, however, between the spacer and the flanking repeats. Sequences with the spacers of two recombining sites are paired at some point in the reaction. The actual binding sites for the FLP protein lie primarily within the flanking repeats.

The evidence for DNA–DNA pairing within the spacers is derived from the properties of a variety of mutations in the spacer, including single base pair insertions, deletions, and point mutations (Senecoff and Cox 1986; Senecoff et al. 1986). Each mutation results in a greatly reduced efficiency of recombination in reactions with unaltered sites. However, the efficiency of the reaction is partially or entirely restored when the reaction is carried out between two sites containing identical spacer alterations (Senecoff and Cox 1986; Senecoff et al. 1986). This implies that the sequence of the spacer is largely irrelevant as long as the two spacers involved in a recombination event are homologous. This homology dependence indicates that DNA–DNA pairing occurs during the reaction. Since no increase in the level of abortive cleavage of the site by FLP protein was observed when sites with nonhomologous spacers were reacted, the important pairing event probably precedes cleavage (Senecoff and Cox 1986).

This result is made more interesting because all of the functional asymmetry of the recombination site lies within the spacer (Senecoff et al. 1986). Asymmetric sites are aligned in the same orientation during recombination, and this determines the outcome or directionality (inversion vs. deletion) of the reaction. We constructed a site that contains a completely symmetrical spacer sequence (TTCTAGAA instead of TCTAGAAA). This involves changes at five of the eight positions in the spacer. The resulting site does not recombine with a site containing an unaltered spacer, but is fully functional in FLP protein-promoted recombination when reacted with a homologous site. This confirms again that the spacer sequence can be altered extensively without affecting recombination as long as the reaction is restricted to sites with homologous spacers. The directionality of the reaction, however, is abolished (Senecoff and Cox 1986). New products are produced that could only arise from a reaction between two sites aligned randomly, in either orientation, during recombination.

The flexibility evident in the spacer sequence implies that this region is not recognized or specifically bound by FLP protein. This is true only for the central 6 bp of the spacer, and mutations at the base pairs at either end of the spacer result in a reduced level of recombination with unaltered sites as described above. Restricting the reaction to two identical sites mutant at this position, however, only resulted in a partial restoration of reaction efficiency (Senecoff and Cox 1986). As will be described below, this position is a specific contact point for the FLP protein.

A variety of evidence indicates that the FLP protein binding sites lie primarily within the 13-bp repeats. The properties of FLP site mutants change abruptly at the boundary between spacer and flanking repeats (Senecoff and Cox 1986). Within the repeats, a reduced level of recombination is observed when mutants and unaltered sites are reacted and this level is reduced still further when two identical mutants are reacted. The properties of the spacer size mutations also implicate the repeats as binding sites. The spacer can be increased or decreased by 1 bp with relatively modest effects on recombination efficiency. Addition of 2 bp in the spacer, however, abolishes recombination (Senecoff and Cox 1986; Senecoff et al. 1986). One possible interpretation of this result is that important protein–protein interactions occur between proteins bound on either side of the spacer. Further, the addition or deletion of 1 bp in the spacer also adds or subtracts 1 bp between the points at which FLP protein cleaves the recombination site (R.C. Bruckner and M.M. Cox, unpubl.). Therefore, as the spacer size is altered, the point at which FLP protein cleaves the site remains fixed relative to the repeats. This again indicates that FLP protein recognizes the repeats rather than the spacer.

The actual binding site has been defined in some detail, as illustrated in Figure 2. The numbering system in Figure 2 originates at the center of the spacer, with the 34 bp of Figure 1 extending from position -17 to $+17$. Specific contact points between FLP protein and this site have been identified in a series of methylation protection and interference experiments (R.C. Bruckner and M.M. Cox,

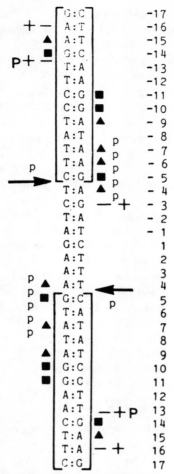

Figure 2 Specific contacts between FLP protein and its recombination site. Symbols: (■) guanine contacts in the major groove; (▲) adenine contacts in the minor groove; (p) phosphate contacts; (+) positions at which purine methylation is enhanced by FLP protein binding; (+P) points at which ethylation of phosphate enhances the efficiency of recombination. Arrows indicate points at which the site is cleaved by FLP protein. The numbering system used in the text is illustrated.

unpubl.). Contacts extend from position 4 to 15 and −4 to −15. Guanine (major groove) and adenine (minor groove) contacts are identified within this region in Figure 2. Most of the contacts lie on one face of the DNA. Phosphate contacts are clustered around the

FLP cleavage sites. An enhancement of methylation resulting from the binding of FLP protein is observed at positions immediately outside the region defined by these contacts. The functional significance of each of the guanine contacts was confirmed in the methylation interference experiments (R.C. Bruckner and M.M. Cox, unpubl.).

The FLP Protein

Purification of the FLP protein to near homogeneity often results in a nearly complete loss of activity. We have found that activity usually can be restored by addition of protein fractions derived from *Escherichia coli* cells that do not produce FLP protein. This enhancement is protease sensitive, and is not duplicated by bovine serum albumin (BSA) or by auxiliary protein factors required in a variety of prokaryotic site-specific recombination systems. We have initiated efforts to purify this protein factor in an attempt to identify it and determine its function. The factor has a large effect on the relative efficiency of inter- and intramolecular recombination reactions and reduces the requirement for FLP protein by as much as two orders of magnitude.

DISCUSSION

The FLP recombination site has now been defined in some detail in terms of its minimal size and the function of its several parts. Characterization of this site, however, is not yet complete. Evidence has been presented that sequences outside the minimal recombination site defined in vitro affect the reaction in vivo (Jayaram 1985). A third 13-bp repeat that lies adjacent to the minimal site is bound and protected by FLP protein in a manner identical to the binding described above (R.C. Bruckner and M.M. Cox, unpubl.). The function of these sequences is unclear.

The availability of a fully characterized recombination site, purified FLP protein, and any additional protein(s) that may affect the activity of FLP protein will permit a complete investigation of the mechanistic details of this recombination event.

REFERENCES
Andrews, B.J., G.A. Proteau, L.G. Beatty, and P.D. Sadowski. 1985. The FLP recombinase of the 2-micron circle DNA of yeast: Interaction with its target sequences. *Cell* **40:** 795.
Babineau, D., D. Vetter, B.J. Andrews, R.M. Gronostajski, G.A. Proteau, L.G. Beatty, and P.D. Sadowski. 1985. The FLP protein of the 2-micron plasmid of yeast. Purification of the protein from *Escherichia coli* cells expressing the cloned FLP gene. *J. Biol. Chem.* **260:** 12313.

Broach, J.R., V.R. Guarascio, and M. Jayaram. 1982. Recombination with the yeast 2μ plasmid is site-specific. *Cell* **29:** 227.

Gronostajski, R.M. and P.D. Sadowski. 1985a. Determination of DNA sequences essential for FLP-mediated recombination by a novel method. *J. Biol. Chem.* **260:** 12320.

———. 1985b. The FLP recombinase of the yeast 2-micron plasmid attaches covalently to DNA via a phosphotyrosyl linkage. *Mol. Cell. Biol.* **5:** 3274.

Hartley, J.L. and J.E. Donelson. 1980. Nucleotide sequence of the yeast plasmid. *Nature* **268:** 860.

Jayaram, M. 1985. Two-micrometer circle site-specific recombination: The minimal substrate and the possible role of flanking sequences. *Proc. Natl. Acad. Sci.* **82:** 5875.

Meyer-Leon, L., J.F. Senecoff, R.C. Bruckner, and M.M. Cox. 1984. Site-specific recombination promoted by the FLP protein of the yeast 2-micron plasmid in vitro. *Cold Spring Harbor Symp. Quant. Biol.* **49:** 797.

Sadowski, P.D., D.D. Lee, B.J. Andrews, D. Babineau, L. Beatty, M.J. Morse, G.A. Proteau, and D. Vetter. 1984. In vitro systems for genetic recombination of the DNAs of bacteriophage T7 and yeast 2-micron circle. *Cold Spring Harbor Symp. Quant. Biol.* **49:** 789.

Senecoff, J.F. and M.M. Cox. 1986. Directionality in FLP protein-promoted site-specific recombination is mediated by DNA-DNA pairing. *J. Biol. Chem.* (in press).

Senecoff, J.F., R.C. Bruckner, and M.M. Cox. 1985. The FLP recombinase of the yeast 2-micron plasmid: characterization of its recombination site. *Proc. Natl. Acad. Sci.* **82:** 7270.

Interaction of the FLP Recombinase of the 2-Micron Plasmid with its Target Sequence

P.D. Sadowski, B.J. Andrews, L.G. Beatty, D. Sidenberg, and G. Proteau
Department of Medical Genetics, University of Toronto
Toronto M5S 1A8 Canada

The 2-micron plasmid of *Saccharomyces cerevisiae* codes for a site-specific recombinase, FLP, that promotes inversion across restricted sequences located within the two 599-bp inverted repeats of the plasmid (Broach et al. 1982). The development of assays that detect the recombination activity of this protein in vitro (Vetter et al. 1983; Meyer-Leon et al. 1984) has permitted the partial purification of the protein (Babineau et al. 1985) and studies of its mechanism of action (Andrews et al. 1985; Gronostajski and Sadowski 1985a,b).

RESULTS

The site of action of the FLP protein is restricted to a region of about 50 bp surrounding an *Xba*I site in the 599-bp inverted repeats (Fig. 1). This region consists of two 13-bp inverted symmetry elements (heavy arrows) surrounding an 8-bp core region. A third 13-bp symmetry element is located in direct orientation with the one to the left of the core. FLP protects about 50 bp from digestion with pancreatic DNase and introduces staggered nicks at the margins of the core region (Andrews et al. 1985). These nicks bear 5'-OH termini, and the 3'-PO_4 groups are covalently attached to a tyrosyl residue of the FLP protein (Gronostajski and Sadowski 1985c). The minimal duplex DNA sequence of this site that is required for in vitro recombination with a wild-type site has been determined to be as small as 22 bp (Gronostajski and Sadowski 1985b). This observation has been confirmed by analysis of completely synthetic FLP sites.

We have studied two mutants of the FLP site that were selected in vivo as being defective in recombination (McLeod et al. 1984). One mutation (624) alters a C/G base pair in the left-hand symmetry element (Fig. 1). This mutation drastically lowers the binding of the

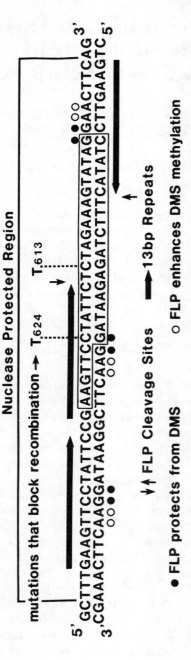

Figure 1 Structure of the FLP recombination site. The heavy arrows represent the three 13-bp symmetry elements. The solid bar above the sequence indicates the region protected by FLP from DNase digestion. The vertical arrows indicate sites of FLP-induced cleavage and the probable sites of strand exchange during recombination. The closed box indicates the minimal sequence required for recombination with a wild-type site. (●) Residues protected by FLP against methylation by dimethylsulfate; (○) enhanced susceptibility to dimethylsulfate. Mutations 624 and 613 decrease FLP-mediated recombination in vivo and in vitro.

FLP protein to the site by changing a critical G residue with which FLP has an important contact in the major groove. A second mutation (613) changes a C/G base pair of the core region to a T/A base pair. This mutant site recombines poorly with a wild-type site but recombines normally with a second site bearing the same change (B.J. Andrews et al., unpubl.). This shows that base-pairing within the 8-bp core region is important for recombination. Similar results have been obtained by Senecoff et al. (1985). From direct studies of other site-specific recombinases (Mizuuchi et al. 1981; Hoess and Abremski 1985) and from in vivo studies in yeast (M. McLeod and J. Broach, pers. comm.), it seems probable that the FLP-mediated crossing over takes place at the margins of the 8-bp core.

The FLP protein forms stable protein–DNA complexes with its target sequence. It is likely that each 13-bp symmetry element constitutes a binding domain for a molecule of FLP. However, truncated sites that undergo efficient recombination are severely impaired in their ability to form stable protein–DNA complexes. We are testing the possibility of cooperative interaction between two FLP sites during recombination.

DISCUSSION

It has been postulated that FLP-mediated recombination promotes amplification of the copy number of the 2-micron plasmid (Futcher 1986). Since we have not been able to uncover any obvious requirement for the third 13-bp symmetry element (extreme left heavy arrow, Fig. 1) in recombination (Andrews et al. 1985), we speculate that this element may play a modulatory role in the recombination event and hence in the determination of plasmid copy number.

ACKNOWLEDGMENTS

This research was supported by the Medical Research Council of Canada.

REFERENCES

Andrews, B.J., G.A. Proteau, L.G. Beatty, and P.D. Sadowski. 1985. The FLP recombinase of the 2μ circle DNA of yeast: Interaction with its target sequences. *Cell* **40**: 795.

Babineau, D., D. Vetter, B.J. Andrews, R.M. Gronostajski, G.A. Proteau, L.G. Beatty, and P.D. Sadowski. 1985. The FLP protein of the 2-micron plasmid of yeast. Purification of the protein from *Escherichia coli* cells expressing the cloned FLP gene. *J. Biol. Chem.* **260**: 12313.

Broach, J.R., V.R. Guarascio, and M. Jayaram. 1982. Recombination with the yeast 2μ plasmid is site-specific. *Cell* **29**: 227.

Futcher, A.B. 1986. Copy number amplification of the 2 μm circle plasmid of *Saccharomyces cerevisiae*. *J. Theor. Biol.* (in press).

Gronostajski, R.M. and P.D. Sadowski. 1985a. The FLP protein of the 2-micron plasmid of yeast. Inter- and intramolecular reactions. *J. Biol. Chem.* **260:** 12328.

———. 1985b. Determination of DNA sequences essential for FLP-mediated recombination by a novel method. *J. Biol. Chem.* **260:** 12320.

———. 1985c. The FLP recombinase of the *Saccharomyces cerevisiae* 2μ plasmid attaches covalently to DNA via a phosphotyrosyl linkage. *Mol. Cell. Biol.* **5:** 3274.

Hoess, R.H. and K. Abremski. 1985. Mechanism of strand cleavage and exchange in the cre-lox site-specific recombination system. *J. Mol. Biol.* **181:** 351.

McLeod, M., F. Volkert, and J. Broach. 1984. Components of the site-specific recombination system encoded by the yeast plasmid 2-micron circle. *Cold Spring Harbor Symp. Quant. Biol.* **49:** 779.

Meyer-Leon, L., J.F. Senecoff, R.C. Bruckner, and M.M. Cox. 1984. Site-specific recombination promoted by the FLP protein of the yeast 2-micron plasmid *in vitro*. *Cold Spring Harbor Symp. Quant. Biol.* **49:** 797.

Mizuuchi, K., R. Weisberg, L. Enquist, M. Mizuuchi, M. Buraczynska, C. Fueller, P.-L. Hsu, W. Ross, and A. Landy. 1981. Structure and function of the phage λ *att* site: Size, int-binding sites, and location of the crossover point. *Cold Spring Harbor Symp. Quant. Biol.* **45:** 429.

Senecoff, J.F., R.C. Bruckner, and M.M. Cox. 1985. The FLP recombinase of the yeast 2 μm plasmid: Characterization of its recombination site. *Proc. Natl. Acad. Sci.* **82:** 7270.

Vetter, D., B.J. Andrews, L. Roberts-Beatty, and P.D. Sadowski. 1983. Site-specific recombination of yeast 2 μm DNA *in vitro*. *Proc. Natl. Acad. Sci.* **80:** 7284.

Site-specific Recombination Promotes Plasmid Amplification in Yeast

F.C. Volkert and J.R. Broach
Department of Molecular Biology
Princeton University, Princeton, New Jersey 08544

The 2-micron circle plasmid of the yeast *Saccharomyces cerevisiae* encodes a specialized recombination system. This system consists of two short homologous sites, designated *FRT*, lying within two 599-bp segments of identical sequence present in inverted orientation within the plasmid, and an enzyme – encoded in the plasmid gene *FLP* – that catalyzes recombination between these two sites (McLeod et al. 1984). Futcher (1986) has proposed that this recombination system provides a mechanism for plasmid amplification, by converting the mode of plasmid replication from theta to rolling circle, through inversion of the relative orientation of the two replication forks at the ends of a replication bubble. In this report we provide experimental confirmation of Futcher's model for amplification.

Two-micron circle is the only stable DNA plasmid found in the baker's yeast *S. cerevisiae*. The plasmid is a 6318-bp circular duplex DNA, with the two inverted repeats bisecting the molecule into nearly equal halves. Since the product of the *FLP* gene actively promotes recombination between these two repeats, 2-micron circle isolated from yeast consists of equal amounts of two distinct isomers that differ in the orientation of one unique region with respect to the other. Although 2-micron circle is present in most wild and laboratory strains at 10^2 copies per cell, no phenotypic change has been associated consistently with its loss. Despite this apparent lack of selection in its favor, and unlike high-copy plasmids constructed from chromosomally derived replicators (*ARS* sequences), 2-micron circle is phenomenally stable. In mitotic growth, it is shed spontaneously from less than 0.01% of cells per generation (Futcher and Cox 1983). Two of the plasmid's coding regions, *REP1* and *REP2*, and a *cis*-acting region of short direct imprecise repeats, *REP3*, are required for equipartitioning, and inactivation of any of these regions reduces 2-micron circle's stability to that of an *ARS* plasmid

(Jayaram et al. 1983). Disruption of either of the remaining coding regions, *FLP* and *D*, or of a *FRT* site, have much less drastic effects on stability, so the biological relevance of these sequences has been an open question (Jayaram et al. 1983).

Recently, Futcher (1986) proposed that *FLP*-mediated recombination mediates copy number amplification in 2-micron circle. Normally, the plasmid's replication is under the same stringent control as that of chromosomal DNA, and density-shift experiments have shown that each plasmid molecule replicates once, semiconservatively, per S phase (Zakian et al. 1979). However, under certain conditions in which the plasmid is introduced into previously [cir⁰] cells at one copy per cell (i.e., by cytoduction), the copy number can rise to its normal level within only a few generations (Sigurdson et al. 1981). Thus, 2-micron circle has the ability to escape cellular control of replication when its copy number is low.

Although the most obvious model for this amplification is reutilization of the plasmid origin during individual S phases (so-called onion skin replication; Botchan et al. 1978), Futcher's model elegantly obviates such generally forbidden multiple initiations. In his proposed scheme, shown in Figure 1, the steps by which amplification occurs are as follows: Semiconservative DNA replication begins at the plasmid origin, progressing bidirectionally (Fig. 1a,b). A *FLP*-mediated recombination reaction occurs during replication (Fig. 1c). This inverts the halves of the circle, reorienting the replication forks so that they travel in the same direction, rather than converging (Fig. 1d). Replication in this mode continues, resulting in a multimeric replication intermediate (Fig. 1e shows a 2.5-mer, but the number of replicative cycles in this mode may vary). Another *FLP* recombination event occurs (Fig. 1f). This restores the original geometry of the replication intermediate, so that the forks once again converge (Fig. 1g). Completion of replication results in a 2-micron circle monomer (Fig. 1i) and a multimer (Fig. 1h, in this case, a dimer). Another *FLP* recombination event (shown as an x), or a general recombination event between homologous sequences, resolves the dimer into two more monomers (Fig. 1j,k), resulting in the production of three monomeric plasmids from a single replication initiation on a monomer. If the time between the two *FLP* inversions (steps shown in Fig. 1c,f) were variable, then any arbitrary number of plasmids could be produced from a single initiation.

Futcher's model predicts that an amplifiable plasmid must have an intact *FLP* system, and that it must undergo *FLP* recombination during its replication. We have tested these predictions using the system shown in Figure 2. Our system comprises three chromo-

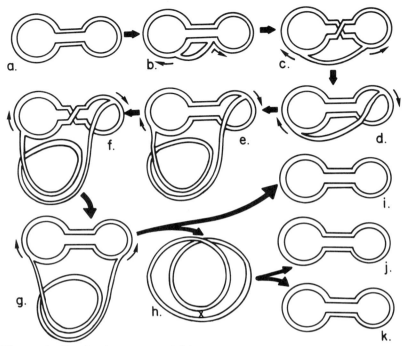

Figure 1 Futcher's (1986) model for recombinational amplification of 2-micron circle. The events shown are described in the text. Arrows indicate the direction of movement of replication forks.

somal insertions: a pair of linearized 2-micron circle flp^- mutants (one of which, FVY5, has both recombination sites intact and one of which, FVY8, has one normal recombination site and one defective recombination site) flanked by directly oriented *FRT* sites, and a *FLP* gene fused to the *GAL10* promoter. Amplification experiments are performed by mating each of the strains with the 2-micron circle insertions with the strain bearing a *GAL10-FLP* insertion, selecting diploids, and then inducing *FLP* production by a shift to galactose medium. The *FLP* protein recombines the *FRT* sites flanking the 2-micron circle inserts, to release one free *flp* Frt$^+$/Frt$^+$ or *flp* Frt$^+$/Frt$^-$ 2-micron circle molecule per cell. *FLP* activity then can invert the former of these, but not the latter. DNA extracted from the diploids is cleaved with restriction enzymes to produce fragments diagnostic of the A and B inversion isomers of 2-micron circle and of a free or chromosomally inserted status of the plasmid sequences. The effect of *FLP* inversion, or failure to invert, on plas-

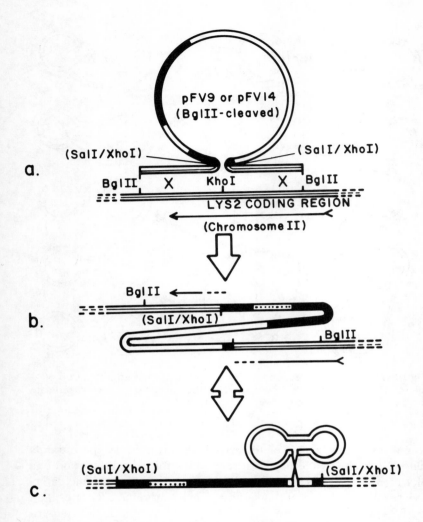

mid copy number is assessed by Southern hybridization, using a probe complementary to 2-micron circle and keeping the intensity of the band derived from the chromosomally inserted *GAL10-FLP* gene constant as a standard.

Results of this analysis, shown in Figure 3, verify the postulated requirement of *FLP* inversion for amplification. Prior to induction of the chromosomal *GAL10-FLP* gene by growth in galactose, there was no evidence of circular, extrachromosomal copies of the 2-micron circle inserts. However, after the shift to galactose medium, both the FVY5 and FVY8 inserts were completely excised from the chromosome. Additionally, the FVY5, but not the FVY8, circular

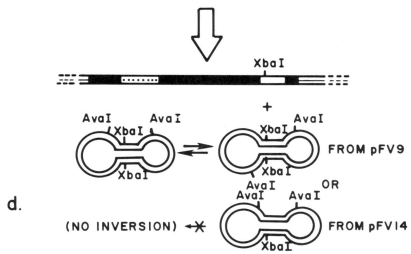

Figure 2 Design of experiments to detect *FLP*-mediated amplification of 2-micron circle. (a) A plasmid bearing a single copy of a *flp* mutant 2-micron circle (▭) has been linearized in sequences homologous to the chromosomal *LYS2* gene (———) and is shown aligned with the chromosomal gene for recombinational insertion. The large Xs show the recombination events leading to insertion. (b) Structure of the same chromosomal region following linear recombinational integration of the plasmid. (c) The same region as in b, redrawn to emphasize the circular recombinational linkage of the 2-micron circle monomer to the other integrated plasmid sequences. (d) The result of exposing this chromosomal region to the *FLP* gene product. *FLP* excises the 2-micron circle moiety from the chromosome. If the moiety's recombination sites are both wild type (the upper excised circle), the excised circle can invert (note *AVA*I sites) and amplify. If one recombination site is inactivated (note missing *Xba*I site in lower excised circle) neither inversion nor amplification occurs. Other symbols: (▰) pBR322 DNA; (······) *URA3* DNA.

excision product was inverted to produce a mixture of A and B isomers, and showed an increase in the order of 10- to 30-fold in copy number, compared with the constant *GAL10-FLP* insert.

Thus, we conclude that invertibility, i.e., the presence of an inverted pair of *FRT* sites, is a requirement for amplification, as predicted. Furthermore, we believe that this requirement reflects a requirement for *FLP* inversion per se, because (1) there is no indication that the *FRT* sites have any function other than to serve as targets for the *FLP* recombinase, and (2) it is unlikely that the mutation we used to inactivate one of them, a four-base insertion, has any effect on distant parts of the plasmid.

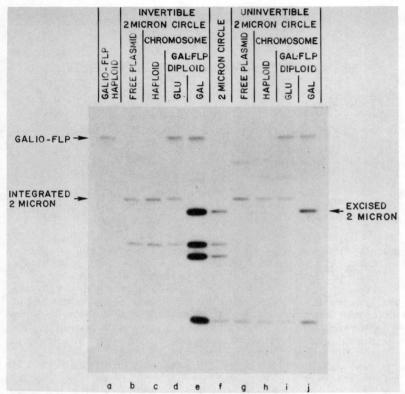

Figure 3 Plasmid amplification promoted by *FLP* protein. The experiment protocol is described in the text. DNA samples were extracted from the indicated strains after growth in glucose or galactose as indicated and cleaved with *Ava*I, fractionated by electrophoresis in 1% agarose, and probed with labeled 2-micron circle DNA. Quantities of DNA loaded on the gel were adjusted to equalize as nearly as possible the intensity of the *GAL10-FLP* fusion band in each lane in which it was present. Using this band as a constant copy-number standard permits one to see changes in the relative levels of the free 2-micron circle bands. (Lane *a*) FVY 2-5A, the haploid strain bearing the integrated *GAL10-FLP* fusion plasmid, pFV17 (which is not cleaved with *Ava*I). (Lane *b*) Purified plasmid DNA of pFV9, the plasmid bearing the invertible *flp* 2-micron circle moiety. (Lane *c*) FVY5, the haploid strain integratively transformed with pFV9. (Lane *d*) The FVY2-5A×FVY5 diploid, grown on glucose. (Lane *e*) The same diploid as in *d*, shifted to galactose. (Lane *f*) Two-micron circle DNA marker. (Lane *g*) Purified plasmid DNA of pFV14, the plasmid bearing the uninvertible *flp* 2-micron circle moiety. (Lane *h*) FVY8, the haploid strain integratively transformed with pFV14. (Lane *i*) The FVY2-5A×FVY8 diploid, grown on glucose. (Lane *j*) The same diploid as in *i*, shifted to galactose.

We have examined the time course of amplification of the invertible and uninvertible plasmids. We found that excision and inversion for both plasmids occurred within 15 minutes of induction of the *GAL10-FLP* gene. For the uninvertible plasmid, no increase in copy number was observed during the next 24 hours of growth. In contrast, as above, the invertible plasmid increased copy number with increasing time. However, there was a lag of more than 4 hours before amplification became apparent, coinciding with a lag in the onset of growth of the culture. This observation is consistent with the model's prediction of a requirement for simultaneous replication and recombination.

A. Murray and J. Szostak (pers. comm.) also have concluded that *FLP* inversion is required for amplification. Their experiments were similar to those just described, except the *FLP* gene used was under control of its normal promoter, and the experiment was begun by a synchronized mating of the strain bearing the invertible (or uninvertible) insertion and the one bearing the *FLP* gene. This method produced a less synchronous onset of recombination, and therefore precluded the demonstration of a temporal lag between recombinational excision and amplification as we found in our experiments. In addition, the uninvertible and invertible inserts were not intact 2-micron circles, but rather in vitro constructions containing one and two copies, respectively, of the 2-micron circle long inverted repeat. Use of these constructs demonstrated that no 2-micron circle genes other than *FLP* were needed for amplification.

Futcher's model explains amplification without recourse to multiple initiations from the plasmid origin in one S phase, and justifies the existence of the *FLP* system, whose adaptive value is implied by the large fraction of the plasmid occupied by the *FLP* gene and the inverted repeats. In addition, Futcher's model explains why all naturally occurring circular DNA plasmids follow a similar organizational scheme (Toh-e et al. 1984; Araki et al. 1985). That is, they all are circular and have a pair of exact inverted repeats which subdivide the remaining unique sequences into nearly equal halves, and they exist in each cell at high copy number as an equal mixture of the two isomers that would be formed by recombinational inversion of the unique segments about the repeats. Finally, Futcher's model rationalizes the relative locations of the 2-micron circle replication origin and the *FLP* recombination sites. These locations maximize the probability that *FLP* inversion during replication will generate the amplifiable intermediate at step c in Figure 1. This step requires that exactly one of the two diverging replication forks has passed a recombination site at the time of recombination, and the chances of this are best when one site is as close to the origin as possible and the other one is as far away as possible.

ACKNOWLEDGMENTS

This work was supported by National Institutes of Health research grant GM 34596 to J.R.B. and Damon Runyon-Walter Winchell postdoctoral fellowship DRG 647 to F.C.V. J.R.B. is an Established Investigator of the American Heart Association.

REFERENCES

Araki, H., A. Jearnpipatkul, H. Tatsumi, T. Sakurai, K. Ushio, T. Muta, and Y. Oshima. 1985. Molecular and functional organizatin of yeast plasmid pSR1. *J. Mol. Biol.* **182:** 191.

Botchan, M., W. Topp, and J. Sambrook. 1978. Studies on simian virus 40 excision from cellular chromosomes. *Cold Spring Harbor Symp. Quant. Biol.* **43:** 709.

Futcher, A.B. 1986. Copy number amplification of the 2 micron circle plasmid of *Saccharomyces cerevisiae*. *J. Theoret. Biol.* (in press).

Futcher, A.B. and B.S. Cox. 1983. Maintenance of the 2 micron circle plasmid in populations of *Saccharomyces cerevisiae*. *J. Bacteriol.* **154:** 612.

Jayaram, M., Y.-Y. Li, and J.R. Broach. 1983. The yeast plasmid 2 micron circle encodes components required for its high copy propagation. *Cell* **34:** 95.

McLeod, M., F.C. Volkert, and J.R. Broach. 1984. Components of the site-specific recombination system encoded by the yeast plasmid 2-micron circle. *Cold Spring Harbor Symp. Quant. Biol.* **49:** 779.

Sigurdson, D.C., M.E. Gaarder, and D.M. Livingston. 1981. Characterization of the transmission during cytoductant formation of the $2\mu m$ DNA plasmid from *Saccharomyces*. *Mol. Gen. Genet.* **183:** 59.

Toh-e, A., H. Araki, I. Utatsu, and Y. Oshima. 1984. Plasmids resembling 2-μm DNA in the osmotolerant yeasts *Saccharomyces bailii* and *Saccharomyces bisporus*. *J. Gen. Microbiol.* **130:** 2527.

Zakian, V.A., B.J. Brewer, and W.L. Fangman. 1979. Replication of each copy of the yeast 2 micron DNA plasmid occurs during the S phase. *Cell* **17:** 923.

Double-strand-break Repair in Yeast Results in Conversion Events That Resemble Mating-type Switching

M. Jayaram

Department of Molecular Biology
Research Institute of Scripps Clinic, La Jolla, California 92037

Gene conversion is the nonreciprocal transfer of information between two allelic genes. In fungi, meiotic gene conversion between homologous chromosomes is frequently associated with recombination of flanking markers (Hurst et al. 1972). Mitotic conversion can also lead to reciprocal exchange of outside markers, but usually less frequently than meiotic conversion (Roman and Jacob 1959). However, unlike conversion between homologous chromosomes, intrachromosomal conversion between repeated loci in yeast is seldom associated with reciprocal exchange (Jackson and Fink 1981; Klein 1984). The switching of mating type in yeast, which involves the replacement of the *MAT* locus by one of the silent loci (*HML* or *HMR*), is formally analogous to a gene conversion event. And, as is the case with other intrachromosomal conversion events, mating-type interconversion is also unaccompanied by recombination of flanking markers (Klar and Strathern 1984).

It has been proposed recently that gene conversion events in yeast can be accommodated by a double-strand-break repair model (Szostak et al. 1983). This model is suggested by the finding that double-stranded breaks in DNA are highly recombinogenic in yeast (Orr-Weaver et al. 1981). A plasmid, gapped within a yeast gene carried on it, can efficiently repair its gap using the chromosomal locus as template. The repair occurs with or without the chromosomal integration of the plasmid, which is equivalent to conversion with or without crossover, respectively (Orr-Weaver and Szostak 1982). Apparently, the initial event in mating-type interconversion is also a double-stranded DNA break caused by the site-specific endonuclease encoded by the *HO* gene (Kostriken et al. 1983). According to the model, the repair of a double-stranded gap by two rounds of single-strand DNA synthesis followed by branch migration would lead to two Holliday junctions. Resolution of the double Holliday junction in the same sense (i.e., cutting inner or outer strands in both cases) results in no crossover of flanking markers; resolution in the opposite sense results in crossover. The absence of crossover during mitotic or meiotic intrachromosomal gene conversion or during mating-type switching may, then, be the consequence of the constraints operating on the resolution of the Holliday junctions.

We describe here experiments which demonstrate that the 2-micron circle recombinase (FLP), a protein capable of making sequence- specific single- or double-stranded cuts in DNA, can initiate a gene conversion event that is reminiscent of mating-type switching. Furthermore, we also demonstrate that a double-stranded gap made in vitro in the vicinity of the FLP-cutting site induces an identical reaction when the gap is allowed to be repaired in vivo. This intrachromosomal conversion, in contrast to mating-type switching, often results in reciprocal crossover irrespective of whether it occurs entirely in vivo or is initiated in vitro and completed in vivo.

Gene Conversion Between the Repeated Segments of 2-Micron Circle Is Associated with Frequent Crossover

The yeast plasmid 2-micron circle contains a 600-bp-long inverted repetition, which we have used as a substrate for gene conversion. The plasmid encodes a site-specific recombinase FLP, which causes crossover within the repeats, resulting in intramolecular inversion. In our experiments we have used a hybrid 2-micron circle plasmid in which the *FLP* gene is mutated, with the consequent elimination of site-specific recombination. The plasmid is gapped in vitro within one of the repeats and the resultant linear molecule is allowed to repair the gap in vivo by using the intact repeat as template. The repair occurs faithfully and efficiently in yeast. We can also assess if the repair results in recombination of flanking markers by determining whether the product of the repair process is the parental plasmid or one that has undergone inversion. The results of such an experiment (diagramed in Fig. 1) demonstrate that double-strand gaps efficiently induce gene conversion in yeast and that the process is reminiscent of conversion between homologous chromosomes in obeying the 50% crossover rule. In experiments not described here (M. Jayaram, unpubl.), we have extended these results to show that the association rule holds regardless of the orientation of the repeats and of the topology of the substrates (circular or linear).

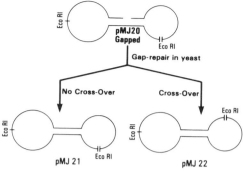

Figure 1 Plasmid pMJ20 contains pBR322 and *LEU2* sequences cloned into the *Eco*RI site within the small unique region of the 2-micron circle genome (not shown in the diagram), thus inactivating the *FLP* gene. The plasmid was linearized by partial digestion with *Xba*I and gapped with BAL-31. The linear molecules were used to transform a $leu2$ [cir^0] yeast strain to leucine prototrophy. Analysis of plasmids in the transformants shows that gap-repair results in plasmid pMJ21 (no crossover) or plasmid pMJ22 (crossover). The parallel lines represent the inverted repeats of the molecule.

A Gene Conversion Event in a 2-Micron Circle::Tn5 Hybrid Plasmid Is Analogous to Mating-type Switching

The cell type in *Saccharomyces cerevisiae* is determined by whether the DNA sequence at the mating-type locus contains **a** or α information. The *MAT***a** and *MAT*α loci have a unique sequence Y**a** and Yα, respectively, bordered by common sequences W-X (to the left) and Z (to the right). When a cell switches its mating type from **a** to α, the net change is the replacement Y**a** at the *MAT* locus by Yα derived from the silent *HML*α information which is located on the same chromosome as the *MAT* locus. The process is apparently initiated by a specific double-stranded cut made within *MAT* at the Y-Z junction by the *HO* enzyme. The rest of the switching reaction fits neatly into the double-strand-break repair model for gene conversion. The conversion, like other intrachromosomal gene conversion events, is not associated with reciprocal crossover, so that, except for the Y**a** to Yα conversion, the chromosome does not undergo any rearrangement. In Figure 2A, I have summarized the sequence of events leading to mating-type switching.

We have constructed a hybrid 2-micron circle plasmid containing a Tn5 insertion within the 2-micron circle small unique region (pcv21::Tn5-36) to test whether a double-stranded break can lead to a conversion event similar to mating-type switching (Fig. 2B). If we

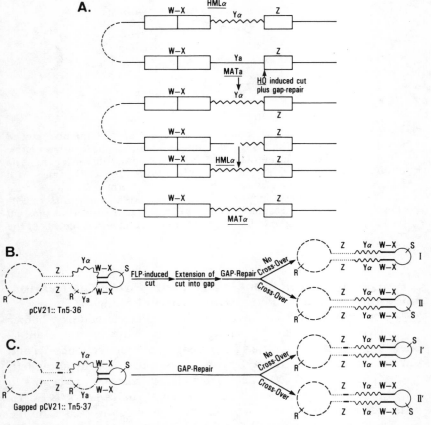

Figure 2 (A) The mating type interconversion from **a** to α is initiated by a double-stranded cut made by the *HO* enzyme at the junction of Y and Z in *MAT*a. The break is presumably extended into a gap and the ensuing gap-repair causes Y**a** to be gene-converted to Yα. The conversion event is not associated with crossover. (R) *Eco*RI; (S) *Sal*I. (B) Plasmid pCV21::Tn5-36, which contains pBR322 and *LEU2* sequences in the *Eco*RI site within the 2-micron circle large unique region (not shown), upon propagation in a [cir⁰] *leu2* yeast strain gives rise to plasmids I and II in which the 2-micron circle segment Y**a** is gene converted to Yα. The conversion occurs with crossover (plasmid I, Tn5 inverted) or without crossover (plasmid II, Tn5 in parental configuration). The conversion is absolutely dependent on *FLP*. (C) pCV21::Tn5-37, linearized by *Xba*I digestion and gapped with BAL-31, repairs the gap in yeast to give plasmids I' and II', which are analogous to I and II, respectively. Plasmids I' and II' differ from I and II in lacking the *Xba*I sites within the 2-micron circle repeats (indicated by ···—····).

call the 2-micron circle repeat "Z," the Tn5 repeats "W-X," and the segments of 2-micron circle unique DNA bordered by the repeats Y**a** and Yα, respectively, we have two mating-type-like loci arranged in inverted orientation on a circular plasmid. We name one of these loci "impostor *MAT***a**" and the other "impostor *HML*" (the silent locus). When the plasmid is propagated in yeast, it gives rise to plasmids I and II in which Y**a** is converted to Yα, that is, the impostor *MAT***a** has now switched to impostor *MAT*α. Since the parent plasmid contains an intact *FLP* gene, which can make a double-stranded break within the 2-micron circle repeat (notice the analogy to *HO*), plasmids I and II can arise by the reaction sequence depicted in Figure 2B, following the rules of the double-stranded-break repair model. We have already established that gap-repair in circular or linear plasmids can occur with or without crossover. Consistent with this rule, plasmid I represents a crossover-associated conversion (Tn5 moiety is inverted), whereas plasmid II exemplifies a noncrossover conversion (Tn5 retains parental configuration).

We derived plasmid pCV21::Tn5-37 from pCV21::Tn5-37 by eliminating the *Xba*I site within one of the 2-micron circle repeats. The repeat lacking the *Xba*I site is immune to breakage by *FLP*. This plasmid was linearized at the *Xba*I site within the second copy of the 2-micron circle repeat (Z of the impostor *MAT* locus) and gapped with BAL-31, and gap-repair was allowed to take place in yeast. The plasmids resulting from the gap-repair have structures analogous to I and II and are named I' and II', respectively. These results clearly demonstrate that a mating-type-like conversion event in a hybrid 2-micron circle plasmid can be mimicked in the absence of *FLP* by making a double-stranded gap in vitro near the normal *FLP* site and carrying out gap-repair in vivo. Since both I' and II' are devoid of the *FLP* sites within the 2-micron circle repeats, neither one of these is capable of undergoing *FLP*-mediated DNA breaks or recombination.

DISCUSSION

We have described here experiments that provide support to the double-strand-break repair model for gene conversion in yeast. These experiments also demonstrate the generality, in yeast, of mating type-like conversion events. Provided the participating loci have the appropriate structural organization, such conversion events can be initiated in vivo by enzymes capable of making double-stranded cuts (for example, *FLP*); or they can also be induced by double-stranded breaks made in vitro. At least two of the site-specific endonucleases from yeast, the 2-micron circle recombinase *FLP* and the *HO* enzyme, have remarkably short recognition sequences. This enables the enzyme sites to be synthesized chemically and inserted into various locations in the yeast genome. Furthermore, it is also possible to obtain regulated expression of these enzymes, using inducible yeast promoters, such as the *GAL10* or *GAL1* promoter. Thus, it is now potentially feasible to explore details of double-stranded-break- induced recombination events as they occur on the yeast chromosomes.

ACKNOWLEDGMENT
My laboratory is supported by funds from the National Institutes of Health.

REFERENCES
Hurst, D.D., S. Fogel, and R.K. Mortimer. 1972. Conversion-associated recombination in yeast. *Proc. Natl. Acad. Sci.* **69**: 101.
Jackson, J.A. and G.R. Fink. 1981. Gene conversion between duplicated genetic elements in yeast. *Nature* **292**: 306.
Klar, A.J.S. and J.N. Strathern. 1984. Resolution of recombination intermediates generated during yeast mating type switching. *Nature* **310**: 744.
Klein, H.L. 1984. Lack of association between intra-chromosomal gene conversion and reciprocal exchange. *Nature* **310**: 748.
Kostriken, R., J.N. Strathern, A.J.S. Klar, J.B. Hicks, and F. Heffron. 1983. A site specific endonuclease essential for mating-type switching in *Saccharomyces cerevisiae*. *Cell* **35**: 167.
Orr-Weaver, T.L. and J.W. Szostak. 1982. Yeast recombination: the association between double-strand gap repair and crossing over. *Proc. Natl. Acad. Sci.* **80**: 4417.
Orr-Weaver, T.L., J.W. Szostak, and R.J. Rothstein. 1981. Yeast transformation: A model system for the study of recombination. *Proc. Natl. Acad. Sci.* **78**: 6354.
Roman, H. and F. Jacob. 1959. A comparison of spontaneous and ultraviolet-induced allelic recombination with reference to the recombination of outside markers. *Cold Spring Harbor Symp. Quant. Biol.* **23**: 155.
Szostak, J.W., T.L. Orr-Weaver, R.J. Rothstein, and F.W. Stahl. 1983. The double-strand break repair model for recombination. *Cell* **33**: 25.

Recombination of Mitochondrial Genes

R.A. Butow, P.S. Perlman,* and A.R. Zinn
Department of Biochemistry
University of Texas Health Science Center at Dallas, Dallas, Texas 75235
*Department of Genetics
The Ohio State University, Columbus, Ohio 43210

In genetic crosses, most yeast mitochondrial DNA (mtDNA) sequences follow a simple input-equals-output rule for their transmission from haploids to diploids: They are quantitatively recovered in the diploid progeny, and the ratio of allelic forms appearing among the products of the cross (bias) is approximately equal (coordinate) for different sequences.

Mitochondrial genomes actively recombine during mating, but overall recombination appears to be limited in that recombination frequencies are only about one-half of the value expected for recombination between distant markers. The molecular (or cellular) basis for this anomalous behavior is unclear. Exceptions to the transmission rule noted above involve two regions of the mitochondrial genome: the 21S rRNA gene and var1, the gene that encodes a protein (var1) associated with the small ribosomal subunit. Certain alleles of these genes are not transmitted coordinately with other mtDNA sequences in crosses, but are selectively lost or retained. Since there is no known selective advantage to cells with any of these alleles, these recombinations can be formally considered as gene conversions. We have studied the molecular basis for the recombination behavior of 21S rRNA and var1 genes, and a number of surprising results have emerged. For var1, we now know that nonreciprocal exchanges are due to the insertion of an optional 46-bp, GC-rich sequence from one allele to another. This GC-rich sequence, called the a element in var1 (Butow et al. 1985), is one of more than 100 related sequences distributed throughout the yeast mitochondrial genome. The distribution of these GC clusters in the mitochondrial genome is highly variable among strains, although the a element is the only such sequence whose recombination has been studied in any detail.

Nonreciprocal exchange at the 21S rRNA locus involves the quantitative insertion of an optional 1.1-kb intron from alleles that have it (omega⁺) to those lacking it (omega⁻). We have demonstrated that expression of an open reading frame (ORF) that encodes a protein of 235 amino acids is required for this process (Macreadie et al. 1985). In contrast, there is no requirement for mitochondrial protein synthesis associated with gene conversion at *var1*. From the analysis of mtDNA in zygotes and the consequences of mutations within the ORF, we concluded that the ORF product initiates the conversion by creating a double-strand break in *omega*⁻ DNA, which is then repaired using *omega*⁺ as a template. This process is reminiscent of the double-strand-break/gap-repair model of Szostak et al. (1983).

Our studies of nonreciprocal exchange at *omega* allowed for the first time a direct determination of the kinetics of mtDNA recombination in zygotes (Zinn and Butow 1984). We examined mtDNA from synchronously mated cells various times after mixing for the presence of a novel restriction fragment arising by recombination between polymorphic alleles; by this assay, recombination was detected as early as 2.5 hours after cells were mixed, essentially concomitant with zygote formation.

The rapidity of *omega* recombination was surprising in light of earlier indirect biochemical measurements which suggested that mtDNA recombination in zygotes might be much slower, requiring perhaps at least 16 hours to be evident (Slonimski et al. 1978; Lopez et al. 1981). We asked whether this difference reflects merely the different nature of the assays, or whether recombination at *omega* might be facilitated by special enzymatic machinery, and in fact be more rapid than recombination elsewhere. To address this question, we constructed strains polymorphic at three loci—*omega*, *var1*, and *cob* (the gene encoding cytochrome *b*)—that would allow us to measure recombination at the DNA level for these loci in the same cross. In contrast to *var1* and *omega*, the *cob* recombination is reciprocal at the population level. Figure 1 shows that recombination initiates at all three loci at the same time (in this cross, 8 hr after cell mixing). Although the diagnostic recombinant forms of *omega* and *var1* appear to increase steadily from this time until 24 hours (the last time assayed), the *cob* recombinant form appears to plateau early.

These data, together with a lower-than-expected overall recombination frequency for unlinked markers, and pedigree studies of Strausberg and Perlman (1978) indicating slow cytoplasmic mixing in zygotes, suggest a physical model for mtDNA recombination in mated cells. Of the 100 or so mtDNA molecules in a zygote, only a fraction of those in the vicinity of the junction formed between the

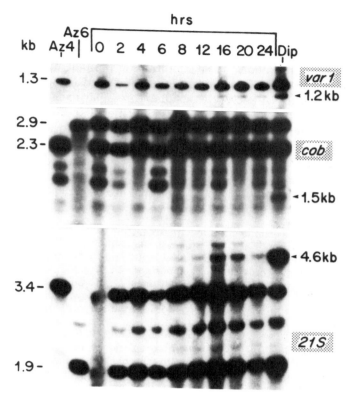

Figure 1 Kinetics of recombination at omega, var1, and cob. Strains AZ4 and AZ6 were synchronously mated and mtDNA was prepared at the indicated times after mixing of cells. DNAs were digested with HpaII (omega), BamHI, and CfoI (var1), or EcoRI, ClaI, and BglII (cob), fractionated on 1% agarose (omega and cob) or 5% acrylamide (var1) gels, transferred to filters, and probed with nick-translated, cloned DNA fragments. The arrows point to the position of fragments derived from nonparental forms of these genes. Only the smaller of the two nonparental reciprocal recombinant cob restriction fragments is shown.

parental cells are mixed and thus free to undergo pairing and recombination. This "zone of mixing" might be constrained by mitochondrial fusion, the position of other organelles in the zygote, or the arrangement of the cytoskeleton. If the number of pairings within this limited pool is large relative to DNA segregation, then reciprocal crossovers will rapidly approach equilibrium, while gene conversion events will continue during the course of mating as they spread through regions outside of this zone of mixing. Given the relatively rapid rate at which mitochondrial genomes segregate, reciprocal recombinations are effectively limited by one or a few productive pairings, whereas at *omega*, and possibly *var1*, directed double-strand breaks would enhance recombinations at these loci. One prediction of this model is that conditions that might alter the availability of parental mitochondria for fusion, and thus change the number of mtDNA molecules available for pairings, would have dramatic effects on recombination frequencies. Experiments of this type are presently under investigation in our laboratories.

ACKNOWLEDGMENTS
This work was supported by grants from the National Institutes of Health and the Robert A. Welch Foundation.

REFERENCES
Butow, R.A., P.S. Perlman, and L.I. Grossman. 1985. The unusual *var1* gene of yeast mitochondrial DNA. *Science* **228**: 1496.
Lopez, I.C., F. Farrelly, and R.A. Butow. 1981. *Trans*-action of the *var1* determinant region on yeast mitochondrial DNA. Specific labeling of mitochondrial translation products in zygotes. *J. Biol. Chem.* **256**: 6496.
Macreadie, I.G., R.M. Scott, A.R. Zinn, and R.A. Butow. 1985. Transposition of an intron in yeast mitochondria requires a protein encoded by that intron. *Cell* **41**: 395.
Slonimski, P.P., P. Pagot, C. Jacq, M. Foucher, G. Perrodin, A. Kochko, and A. Lamouroux. 1978. Mosaic organization and expression of the mitochondrial DNA region controlling cytochrome *c* reductase and oxidase. I. Genetic, physical, and complementation maps of the *box* region. In *Biochemistry and genetics of yeast: Pure and applied aspects* (ed. M. Bacila et al.), p. 339. Academic Press, New York.
Strausberg, R.L. and P.S. Perlman. 1978. The effect of zygotic bud position on the transmission of mitochondrial genes in *Saccharomyces cerevisiae*. *Mol. Gen. Genet.* **163**: 131.
Szostak, J.W., T.L. Orr-Weaver, R.J. Rothstein, and F.W. Stahl. 1983. The double-strand-break repair model for recombination. *Cell* **33**: 25.
Zinn, A.R. and R.A. Butow. 1984. Kinetics and intermediates of yeast mitochondrial DNA recombination. *Cold Spring Harbor Symp. Quant. Biol.* **49**: 115.

A Recombination-stimulating Sequence in the Ribosomal RNA Gene Cluster of Yeast

G.S. Roeder, R.L. Keil, and K.A. Voelkel-Meiman
Department of Biology, Yale University, New Haven, Connecticut 06511

The ribosomal RNA (rRNA) genes of yeast are clustered on a 9-kbp segment of DNA. As indicated in Figure 1, the 18S, 5.8S, and 25S rRNA genes are transcribed as a 35S precursor RNA and the 5S gene is transcribed separately from the opposite strand (Bell et al. 1977; Philippsen et al. 1978). This cluster of four genes is present in approximately 140 copies per haploid genome and these repeats are organized in a single, tandem array on chromosome XII (Schweizer et al. 1969; Petes 1979).

One very striking feature of the rDNA is that all the repeat units within an array are identical, or nearly identical, to each other (Petes et al. 1978). One mechanism that may be responsible for this sequence homogeneity is genetic recombination. Gene conversion and crossing-over between repeats can lead to the correction or loss of new mutations (Szostak and Wu 1980). It has been suggested that special mechanisms for promoting recombination between repeat

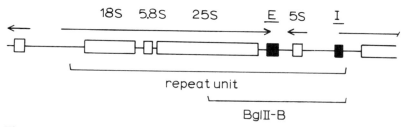

Figure 1 The structure of the rRNA gene cluster of yeast. The open boxes indicate the coding sequences for the mature rRNAs. The arrows above the line indicate transcription units; the arrowheads indicate the direction of transcription. The brackets below the line indicate the limits of one repeat unit and the position of the BglII-B fragment. The black boxes labeled E and I indicate the initiator and enhancer fragments required for the stimulation of recombination.

units might exist to maintain sequence homogeneity in certain multigene families, such as the rDNA (Gojobori and Nei 1984). In this paper, we demonstrate that the rDNA of yeast displays recombination-stimulating activity.

Plasmid Recombination

We have demonstrated that the BglII-B fragment (see Fig. 1) of rDNA has the ability to stimulate recombination between mutant *HIS4* genes carried by an autonomously replicating yeast plasmid (Keil and Roeder 1984). In this assay, the BglII-A fragment of rDNA, the other half of the repeat unit, displays no recombination-enhancing activity. In addition, BglII fragments representing approximately 35% of the yeast genome were screened and none was capable of stimulating intraplasmid exchange. We have designated the recombination-stimulating sequence present in the BglII-B fragment of rDNA as *HOT1*.

Interchromosomal Recombination

We have determined the effect of *HOT1* on interchromosomal recombination at both the *HIS4* and *LEU2* loci. These experiments utilized diploid strains carrying *his4* or *leu2* heteroalleles. The BglII-B fragment of rDNA was inserted to one side of the recombining gene on both copies of the homologous chromosome. *HOT1* leads to a seven- to eightfold increase in the frequency of His$^+$ or Leu$^+$ prototrophs resulting from interchromosomal exchange. The activity of the BglII-B fragment of rDNA is orientation dependent; it stimulates prototroph formation only when the 5S gene is proximal to, and the 25S gene is distal to, the recombining genes. Mitotic, but not meiotic, recombination frequencies are affected by *HOT1*. A stimulation of interchromosomal recombination is observed only when *HOT1* is present on both copies of the chromosome.

HIS4 and *LEU2* are both located on the left arm of chromosome III, about 25 kbp away from each other. When *HOT1* is present at either *HIS4* or *LEU2*, it does not affect recombination at the other locus.

Intrachromosomal Recombination

To examine the effect of *HOT1* on intrachromosomal recombination, we have constructed a strain in which chromosome III carries a duplication of *his4* sequences, as indicated in Figure 2. This strain carries one complete copy of the *HIS4* gene, which is His$^-$ due to the presence of an amber (*his4-260*) mutation. In addition, the chromosome carries one incomplete *his4* gene, which is wild type in sequence and includes the site of the *his4-260* mutation. Inserted

Figure 2 Construction used to assay intrachromosomal recombination at *HIS4*. The solid line indicates the vector sequences which consist of pBR322 and the yeast *URA3* gene. The open arrow on the right indicates the *HIS4* gene; * indicates the *his4-260* mutation. The open box to the left represents an incomplete *HIS4* gene containing wild-type sequence information. The dashed lines indicate chromosome III DNA upstream of the *HIS4* coding region. The parentheses below the line indicate those *HIS4* and *HIS4*-adjacent sequences that are present in two copies; one on each side of the inserted vector. The *Bam*HI site in the left repeat is the site at which the *Bgl*II-B fragment of rDNA, and subclones thereof, were inserted to examine their effects on the frequencies of His⁺ and Ura⁻ recombinants.

between the *HIS4* repeats are pBR322 vector sequences and the yeast *URA3* gene. In strains carrying this construction, recombination can be measured in two different ways. First, His⁺ prototrophs can be selected; these result from gene conversion events in which the *his4-260* mutation is converted to wild type using the information present in the incomplete copy of the gene. Second, reciprocal crossovers that lead to excision of the plasmid vector and loss of the duplication can be selected on medium containing 5-fluoro-orotic acid, which allows the growth only of Ura⁻ cells. These Ura⁻ recombinants include both His⁺ and His⁻ derivatives; the phenotype depends on whether or not the *his4-260* mutation remains on the chromosome following plasmid excision.

In order to examine the effect of *HOT1* on intrachromosomal recombination at *HIS4*, we have inserted the *Bgl*II-B fragment of rDNA at the *Bam*HI site present in the left copy of the *HIS4* duplicated sequences (see Fig. 2). In strains carrying the rDNA at this site, the frequency of His⁺ and Ura⁻ recombinants is 1.1×10^{-3} and 1.7×10^{-3}, respectively. When no rDNA is present at *HIS4*, these frequencies are approximately 10-fold lower.

In the assay of intrachromosomal recombination, as for interchromosomal recombination, *HOT1* shows an absolute orientation dependence. The frequencies of His⁺ and Ura⁻ recombinants are increased only when the 5S gene lies to the right, and the 25S gene to the left, in the construction shown in Figure 2.

Localization of HOT1

To define the sequences required for *HOT1*-stimulated recombination, we have constructed and analyzed a variety of subclones and deletion mutants of the *Bgl*II-B fragment. These subclones were inserted at the *Bam*HI site indicated in Figure 2 and their effects on the frequency of His$^+$ and Ura$^-$ recombinants were determined. These experiments indicate that two noncontiguous segments of DNA are required for the stimulation of exchange. One of these is a 252-bp *Eco*RI–*Sma*I restriction fragment located near the right end of the fragment (see Fig. 1). The other is a 318-bp *Eco*RI–*Hpa*I fragment located near the middle of the fragment (see Fig. 1). When a subclone containing only these two small fragments is inserted at *HIS4*, the frequency of His$^+$ prototrophs is increased approximately 25-fold and the frequency of Ura$^-$ recombinants is increased about 100-fold relative to the frequencies observed when no rDNA is present at *HIS4*. It is interesting that the 570-bp subclone stimulates recombination to a significantly greater extent than the intact *Bgl*II-B fragment.

Elion and Warner (1984) have identified the sequences required for efficient transcription of the 35S precursor rRNA. The sequences shown by these workers to be required for 35S transcription initiation are the same sequences required for the stimulation of recombination. The *Eco*RI–*Sma*I fragment near the right end of the *Bgl*II-B fragment contains the transcription initiation site. Contained within the *Eco*RI–*Hpa*I fragment is a *Eco*RI–*Hin*dIII fragment shown by Elion and Warner to contain an enhancer of RNA polymerase I transcription. This enhancer stimulates transcription at the 35S initiation site approximately 15-fold. This coincidence between sequences required for transcription and those involved in stimulating recombination suggests that transcription initiating in the rDNA and proceeding through the adjacent genes may be responsible for the enhanced genetic exchange. This theory is consistent with the orientation dependence of *HOT1*. The orientation of the fragment that is active in stimulating recombination is that in which transcription initiating in the rRNA would proceed toward the adjacent genes.

Orientation Dependence
To determine whether the enhancer and/or the initiator fragment is responsible for the directionality of *HOT1*, we have constructed strains in which one or both of these fragments is inverted relative to the adjacent *HIS4* genes. The frequencies of His$^+$ and Ura$^-$ intrachromosomal recombinants were then determined using the assay system diagramed in Figure 2. These experiments indicate that the initiator fragment functions only in one orientation, that in which rDNA-promoted transcription proceeds toward the *HIS4*

genes. However, the enhancer functions in either orientation. In all these constructions, the enhancer fragment lies upstream of the initiator (i.e., to the left in Fig. 2).

DISCUSSION

The experiments just described indicate that transcription initiating in the rDNA and proceeding through the adjacent gene may be responsible for *HOT1*-stimulated exchange. It appears that *HOT1* must act on both copies of the homologous sequence for recombination to be stimulated. It is possible that transcription by RNA polymerase I leads to a localized unwinding of DNA. If so, then exposed single strands from homologous repeats could engage in complementary base-pairing. This pairing, or synapsis, might be the first step in *HOT1*-promoted genetic exchange.

REFERENCES

Bell, G.I., L.J. DeGennaro, D.H. Gelfand, R.J. Bishop, P. Valenzuela, and W.J. Rutter. 1977. Ribosomal RNA genes of *Saccharomyces cerevisiae*. I. Physical map of the repeating unit and location of the regions coding for 5S, 5.8S, 18S, and 25S ribosomal RNAs. *J. Biol. Chem.* **252**: 8118.

Elion, E.A. and J.R. Warner. 1984. The major promoter of rRNA transcription in yeast lies 2 kb upstream. *Cell* **39**: 663.

Gojobori, T. and M. Nei. 1984. Concerted evolution of the immunoglobulin V_H gene family. *Mol. Biol. Evol.* **1**: 195.

Keil, R.L. and G.S. Roeder. 1984. *Cis*-acting recombination-stimulating activity in a fragment of the ribosomal DNA of *S. cerevisiae*. *Cell* **39**: 377.

Petes, T.D. 1979. Meiotic mapping of yeast ribosomal deoxyribonucleic acid on chromosome XII. *J. Bacteriol.* **138**: 185.

Petes, T.D., L.M. Hereford, and K.G. Skryabin. 1978. Characterization of two types of yeast ribosomal RNA genes. *J. Bacteriol.* **134**: 295.

Philippsen, P., M.J. Thomas, R.A. Kramer, and R.W. Davis. 1978. Unique arrangement of coding sequences for 5S, 5.8S, 18S and 25S ribosomal RNA in *Saccharomyces cerevisiae* as determined by R-loop and hybridization analysis. *J. Mol. Biol.* **123**: 387.

Schweizer, E., C. MacKechnie, and H.O. Halvorson. 1969. The redundancy of ribosomal and transfer RNA genes in *Saccharomyces cerevisiae*. *J. Mol. Biol.* **40**: 261.

Szostak, J.W. and R. Wu. 1980. Unequal crossing over in the ribosomal DNA of *Saccharomyces cerevisiae*. *Nature* **284**: 426.

Switching Genes in *Schizosaccharomyces pombe* and DNA Rearrangements in the Mating-type Region of *S. pombe*

H. Gutz, L. Heim, P. Kapitza, and H. Schmidt

Institut für Genetik, Technische Universität Braunschweig
D-3300 Braunschweig, Federal Republic of Germany

Homothallic (h^{90}) strains of *Schizosaccharomyces pombe* have three mating-type (MT) genes that are closely linked. They map in the MT region on chromosome II in the order *mat1-P,M–L–mat2-P-K–mat3-M*. *mat1* is the site of gene expression, where either *P*lus or *M*inus information resides, and *mat2* and *mat3* are silent cassettes. The intervening regions are called L (16.6 kb) and K (15 kb). In h^{90} strains, the cells frequently switch their MTs by transposing gene copies from the cassette genes into *mat1*. Heterothallic strains (h^+ or h^-) originate by duplications or deletions in the MT region. The loss of the L region is lethal, and h^{+L} and h^{-L} deletions are known (Beach 1983; Beach and Klar 1984).

In h^{90} strains, mutations occur that reduce the frequency of MT switching. They map either in any of 10 different *swi* genes or in a switching signal, *smt*, at *mat1* (Egel et al. 1984; Gutz and Schmidt 1985). The lethal deletions in the MT region can be complemented by *mat2:1°* plasmids (Gutz and Egel 1984). Here we report on (1) different properties of *swi* mutants (e.g., radiation sensitivity, influence on recombination) and (2) various reintegration patterns of *mat2:1°* plasmids into the MT region.

Switching Genes

The 10 hitherto known *swi* genes are listed in Table 1. Since MT switching is a process akin to recombination and DNA repair, we tested the *swi* mutations for radiation sensitivity and for an influence on recombination frequencies. Six *swi* genes turned out to increase the sensitivity for UV and/or γ-rays (Table 1) (part of these experiments were made in collaboration with A. Nasim, National Research Council of Canada, Ottawa). Because of the above result, it appeared likely that some of the 22 *rad* genes described in *S.*

Table 1 Radiation sensitivity of *swi* mutants

Class[a]	Gene	UV sensitivity[b]	γ-Ray sensitivity[b,c]
Ia	swi1	(+)	+
	swi3	−	−
	swi7	−	(+)
Ib	swi2	−	−
	swi5	+	+
	swi6	−	+
II	swi4	−	−
	swi8	−	−
	swi9	+	−
	swi10	+	+

[a] h^{90} *swi* mutants of class II (but not those of class I) yield h^+, h^-, and sterile segregants; in the class Ia mutants the frequency of double-stranded breaks at *smt* is reduced (Egel et al. 1984; Gutz and Schmidt 1985).
[b] (+) increased sensitivity; (−) sensitivity not increased.
[c] Anwar Nasim, pers. comm.

pombe (Phipps et al. 1985) might also reduce MT switching. We found this to be true for *rad10*, *rad16*, and *rad20*. In h^{90} strains, these mutations act like *swi* genes of class II. However, the three *rad* genes are not allelic with any of the *swi* genes (*swi9* has not yet been tested).

In other experiments we examined all *swi* genes except *swi6* and *swi8* for a possible influence on meiotic intragenic recombination in the *ade6* locus. Only *swi5* causes a significant reduction of the recombination frequencies. This gene also reduces the frequencies of gene conversion (*ade6-M26* was tested) and of crossings-over between *his7* and *his2*. The latter two markers flank the MT region.

In the *swi* mutants, MT switching still takes place, albeit with reduced frequencies. Therefore, we tested whether a cumulative effect occurs if two different *swi* genes are present in an h^{90} strain. By appropriate crosses, we prepared 45 strains containing the *swi* genes in all possible pairwise combinations. All strains in which *swi* genes of different classes (classes Ia and Ib are considered as different) were combined showed a significant reduction of MT switching as compared with the single gene mutants. Such a cumulative effect was not observed in the combinations that included *swi* genes of the same class. As to the class Ia genes, however, the combinations *swi1 swi7* and *swi3 swi7* are lethal: Ascospores of these two genotypes still germinate and form microcolonies, but then the growth of the colonies ceases.

DNA Rearrangements in the Mating-type Region

From diploid h^{90}/h^{-S} strains, h^{+L}/h^{-S} mutants can easily be isolated. The latter produce azygotic asci with two viable h^{-S} and two dead spores. However, in about 2% of the asci a third spore germinates and forms a colony that shows a weak iodine reaction because of the presence of few two-spored azygotic asci. For that reason we call the spore clones wip (weakly iodine positive). They have an h^+ mating type. Several experiments indicated that the wip clones still contain the h^{+L} deletion but possess an extrachromosomal element that can complement this deletion (Gutz and Egel 1984). We supposed the latter element to be the mat2:1° plasmid described by Beach and Klar (1984) for h^{-S} strains. This plasmid consists of the L region, a mat2:1 cassette with Minus information, and the smt signal.

The first wip clones are now called wipI since a second type, wipII, was detected subsequently (see below). Hybridization experiments with chromosomal DNA from wipI clones confirmed the assumptions stated in the preceding paragraph. The wipI clones are mitotically and meiotically unstable: Due to losses of the plasmid, they produce many nonviable cells and ascospores, respectively.

A tetrad from an h^{+L}/h^{-S} strain yielded two nonviable spores and two spore clones of a new phenotype, wipII. Colonies of wipII strains also exhibit a weak iodine reaction that, however, is more intensive than that of wipI colonies. wipII strains contain zygotic asci; they do not produce nonviable cells or ascospores. DNA hybridization experiments showed the presence of an h^{-S} MT region and of mat2:1° plasmids with Plus information. Evidently, the cassette on the plasmid can switch from M to P and vice versa if proper donor cassettes are present in the chromosomal MT region(s). Such switchings in mat2:1° take place in wipI as well as in wipII strains.

The wipI clones segregate h^{90} and h^+ strains; occasionally also h^- strains are found. These segregants no longer give rise to nonviable cells or ascospores. They originate by reintegration of the plasmids into the chromosomal MT region whereby various DNA rearrangements are created (Fig. 1). It should be noted that the segregants of type d only can arise if a plasmid with P information integrates in mat3:3 by a crossing-over in the proximal homology boxes. The occurrence of these segregants is an additional proof of the presence of P plasmids in the wipI strains.

In wipII strains, which originate from h^{-S} cells with a mat2:1°(P) plasmid, the plasmid can also integrate in various ways into the chromosome. The resulting MT regions have two M cassettes, one

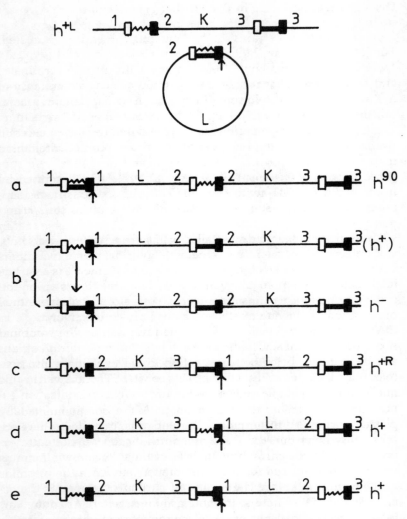

Figure 1 (*a–e*) Segregants of *wipI* strains that arise by different methods of plasmid reintegration into the h^{+L} mating-type region. In the cassettes, *P* information is symbolized by ～ and *M* information by −; when both symbols are shown, either *P* or *M* can reside because of MT switching. The short arrows (↑) indicate the *smt* signal. For cassette type numbering and other symbols, see Beach and Klar (1984). In *a*, the h^{90} region is restored from that which h^{+L} originally had arisen; *c* corresponds to an h^{+R} strain. Segregants of types *a* and *c* are the most common. Segregants *b* and *d* are less frequent, and they have configurations that were not known until now. Segregants of type *e*, which also should occur, have not yet been found. (The figure is not drawn to scale.)

P cassette, and two L regions. Such rearrangements were not known up to the present. The *wipII* strains give rise to h^{-s} segregants.

DISCUSSION

Our results show that part of the *swi* genes play a role not only in MT switching but also in repair of radiation damage and in recombination. On the other hand, three *rad* genes reduce MT switching. These observations open in *S. pombe* the possibility of elucidating which mechanisms are common and which are specific for the three phenomena.

According to Egel et al. (1984), three different steps are involved in the process of MT switching. The *swi* genes of class Ia are supposed to be necessary for step 1, those of class Ib for step 2, and those of class II for the third and final step. It is remarkable that in strains with two mutated *swi* genes a cumulative effect occurs only if the genes belong to different classes, i.e., if defects in two different steps are present. If the genes belong to the same class and thus affect only one step, it seems not to matter whether one or two gene products are defective. Apparently for each of the steps a residual background activity exists that is not eliminated by mutations in the respective *swi* genes. However, the findings that the combinations *swi1 swi7* and *swi3 swi7* are lethal pose an important question: Are these genes, besides their function in MT switching, also necessary for some other functions that are vital? The answer to this question remains to be determined in further experiments.

In the *wipI* clones, the lethal deletions are complemented by *mat2:1°* plasmids. Furthermore, sporulation is promoted. These observations show that the L region and the *mat2:1* cassette are expressed on the plasmids. The analysis of the *wipI* and *wipII* strains revealed frequent and manifold interactions between the plasmids and the chromosomal MT regions. Of particular interest is the occurrence of MT switching in the plasmid cassettes. As to the various integration patterns of the plasmids into the chromosome, in most cases it cannot be decided whether these result from switching events accompanied by crossings-over, or from crossings-over alone. For the configurations b, d, and e in Figure 1, this ambiguity does not exist; these configurations only can arise by crossings-over in the homology boxes without a switching event.

In MT switching, double-stranded breaks at *smt* are involved. Such breaks are also postulated by Szostak et al. (1983) for gene conversion. The *mat2:1°* plasmids possess a *smt* signal and thus exhibit a distinct recombinational activity. They may prove to be useful tools for studies on recombination, especially in combination with *swi* or *smt* mutations.

ACKNOWLEDGMENT
Our work was supported by Deutsche Forschungsgemeinschaft.

REFERENCES
Beach, D.H. 1983. Cell type switching by DNA transposition in fission yeast. *Nature* **305**: 683.

Beach, D.H. and A.J.S. Klar. 1984. Rearrangements of the transposable mating-type cassettes of fission yeast. *EMBO J.* **3**: 603.

Egel, R., D.H. Beach, and A.J.S. Klar. 1984. Genes required for initiation and resolution steps of mating-type switching in fission yeast. *Proc. Natl. Acad. Sci.* **81**: 3481.

Gutz, H. and R. Egel. 1984. Lethal mutations in the mating-type region of *Schizosaccharomyces pombe*: Rescue by extrachromosomal elements. In *Abstract from 12th International Conference on Yeast Genetics and Molecular Biology*, Edinburgh, p. 73.

Gutz, H. and H. Schmidt. 1985. Switching genes in *Schizosaccharomyces pombe*. *Curr. Genet.* **9**: 325.

Phipps, J., A. Nasim, and D.R. Miller. 1985. Recovery, repair, and mutagenesis in *Schizosaccharomyces pombe*. *Adv. Genet.* **23**: 1.

Szostak, J.W., T.L. Orr-Weaver, R.J. Rothstein, and F.W. Stahl. 1983. The double-strand-break repair model for recombination. *Cell* **33**: 22.

Relationship of DNA Breakage at the *smt* Site and Mating-type Switching in *Schizosaccharomyces pombe*

R. Egel

Institute of Genetics, University of Copenhagen
DK-1353 Copenhagen K, Denmark

Mating-type expression and mating-type switching in yeasts have been worked out in amazing detail as an example of a master switch control pivotally placed at a multiple branch point of cellular regulation. The homothallic interconversion of mating types, in particular, has been shown to be mediated by the transposition of defined DNA segments, the cassettes, from silent storage loci to a unique expression site. In general terms the two species of yeast best analyzed, *Saccharomyces cerevisiae* and *Schizosaccharomyces pombe*, show basic similarities, in spite of their wide phylogenetic separation. In more detail, however, many significant differences can be observed, and the continued comparison of the underlying mechanisms in both yeasts will further our understanding of how such intricate systems of gene activation might have evolved.

Mating-type Cassettes in *S. pombe*

All the cassettes participating in mating-type switching in *S. pombe* have been cloned and characterized molecularly (Beach 1983) and their DNA has been sequenced (M. Smith and J. Burke, pers. comm.). Basically there are two different cassettes, one for each mating type, *P* for "plus" and *M* for "minus," and these are found in three different surroundings: *mat1*, which is the expression site for either *P* or *M* information, *mat2*, which only stores *P*, and *mat3*, which only stores *M* information. All three cassettes are linked genetically as direct repeats on the right arm of chromosome II. Their internal sequences are different over 1.1 kb, and these differential segments are flanked by boxes of homology, which are identical at all three loci (or nearly so at two of them): *H1* of 59 bp is located to the right, and *H2* of 135 bp is located to the left at all three loci, whereas *H3* with 51 (of 55) identical base pairs is located to the left of *H2* at the two silent loci *mat2* and *mat3* exclusively.

Double-strand Cuts at the *smt* Signal

The *smt-s* mutation defines a *cis*-acting switching signal mapping to the right of *mat1* (Egel and Gutz 1981). It drastically reduces the frequency of mating-type switching, and it has since been characterized as a small deletion in the vicinity of the *H1* box (Beach 1983). Close to that site a high proportion of DNA molecules extracted from wild-type cells are broken by a double-strand cut, but not so in the mutant. The relative frequency of broken molecules does not vary much during the cell cycle, as analyzed by starvation or *cdc* arrest in various mutants (Beach 1983) or in a synchronous culture (O. Nielsen, unpubl.). On the other hand, mutations in various *swi* genes, which also lead to a reduced frequency of mating-type switching, may or may not affect the level of *smt* breakage (Egel et al. 1984). In double mutants, the absence of cuts was epistatic over their presence. Hence the respective gene products were assumed to participate in a sequential pathway concerned with the introduction of cuts at *smt*, the initiation of mating-type switching from these cuts (by copying a silent cassette), and the proper resolution of the resulting intermediate at the left cassette boundary (*H2*). The resolution step is differentially affected in a peculiar subclass of *swi* mutants, which reach normal levels of cutting but give rise to aberrant switch results by copying and inserting too-long segments, containing both silent cassettes together with the entire spacer region in between.

Chromosome Imprinting for High Switching Potential

As discovered by Miyata and Miyata (1981), mitotic division in the haploid homothallic strain of *S. pombe* frequently results in the asymmetric segregation of switching potential. In minipedigrees the appearance of a sister-cell zygote is evidence of one switching event in the preceding cell division, but (next to such a zygote) conjugation never did occur in the other pair of sisters in a chain of four related cells. Therefore, the founder cell of such a clone of four divided asymmetrically. Whenever one of its daughters switched mating type in the next division, the other daughter was unable to do so simultaneously. Such asymmetric segregation of switching potential was also observed in pedigrees of diploid cells where two chromosomes could switch independently (Egel 1984), but there the asymmetry was limited to each chromosome individually, i.e., a switch of mating type in a dividing cell on one of its chromosomes did not preclude the sister cell from switching mating type on the other chromosome. On the contrary, under these conditions a switch in the other cell was about twice as likely as compared with the average in the entire population.

In another series of diploid pedigrees designed to test mating-type switching on one chromosome while the other was fixed in the heterothallic (+) or (−) configuration (R. Egel and B. Eie, unpubl.), it appeared that recurrent switches were overabundant in the following sense. A mating-type switch in one cell division results in a mixed pair of sister cells. As compared to the mother cell, one of the daughters expresses the opposite mating type whereas the mother's mating type is conserved in the other daughter. In the next division the daughter with the conserved mating type gave rise to another mixed pair (by switching mating type again) about twice as frequently as the average population. The daughter with the opposite mating type terminated the pedigree by sporulation and thus could not be followed any further. Yet, it can be concluded that a virgin cell just having emerged from a switch to the opposite mating type should always begin with a low probability to switch back to the original mating type. Otherwise Miyata's rule would frequently be violated.

Toward a Mechanism of Cutting at *smt*

In *S. cerevisiae* the initial cut at the *MAT* locus is made by the *HO* endonuclease, but attempts to isolate an equivalent enzyme activity from *S. pombe* have not yet been successful. Is the mechanism entirely different, as the stability of the cuts throughout the cell cycle and the asymmetric segregation of switching potential by chromosome imprinting might suggest? The participation of an undetermined nuclease is nevertheless supported by the following in scriptu analysis of the DNA sequences in the vicinity of the *smt* signal. The YZ endonuclease encoded by the *HO* gene shares some residual homology with the consensus recognition sequence of two other nucleases from *S. cerevisiae*, *ScaI* and *ScaII* (Shibata et al. 1984). When the consensus sequence for *ScaI* came out, I compared it to the entire *mat P* and *M* cassettes of *S. pombe* in both directions. The best-matching fit showed up at the *mat/H1* boundary, just where the *smt* cuts were expected to map (their exact position is still not known), and the fit was a little better toward the inside of the cassette (in contrast to *S. cerevisiae*). A full comparison of the respective sequences is displayed in Figure 1. The sequences are partially symmetrical as "dirty palindromes." Embedded in this symmetry, considerable residual homology to the essentially asymmetric consensus sequences from *S. cerevisiae* is found in either direction. The

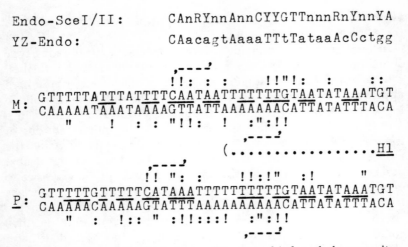

Figure 1 DNA homologies between known and inferred cleavage sites. The reference sequences are the *SceI*/II consensus and the *MAT*-YZ boundary from *S. cerevisiae*, to be compared with the *S. pombe* sequences from the *mat M*/*H1* and *P*/*H1* boundaries. In the reference sequences, the consensus is indicated by capital letters (R = G or A, Y = C or T). In the *S. pombe* sequences, palindromic symmetry is underlined. The matching bases fit to both references (!), *Sce* only ("), or YZ only (:). The staggered cuts as occurring in *S. cerevisiae* are indicated above and below the corresponding match symbols (,----').

corresponding cleavage points, which in *S. cerevisiae* are staggered by four bases, are indicated. If one of these sites is used in *S. pombe*, it should be the right one (belonging to the recognition site extending into the cassette), as only those cuts would lie in the *H1* box common to both mating types. It is interesting to note that the consensus fit is slightly better for *P* than for *M*, and it is *P* that usually switches to *M* more frequently than vice versa.

DISCUSSION

It is now firmly established that a double-strand cut has a crucial function in mating-type switching in *S. pombe*. It is most reasonable to assume that these cuts are introduced during S phase in one branch of the replication fork passing the *smt* signal. During the next S phase in one of the branches of the new replication fork, the cut is healed by the insertion of a new copy from a silent cassette (using the exposed 3' end as the first primer). On the other branch a cut seems usually to be retained for another cell cycle, as suggested by the high frequency of recurrent switches in the same direction for several divisions in a row. In this model the cell determined to switch in the next division is a G_1 cell carrying a cut at *smt*, and the cut itself provides the molecular basis for the inferred chromosome imprinting, at least during most of the cell cycle preceding a mating-type switch. It is still an open question as to how such a long-lasting DNA cut might be stabilized. One is tempted to speculate that some supportive protein structure bridging the gap should be essential, in which the products of the many *swi* genes might be incorporated.

ACKNOWLEDGMENT
The generous communication of sequence data before publication by Dr. M. Smith (Vancouver) is gratefully acknowledged.

REFERENCES

Beach, D.H. 1983. Cell type switching by DNA transposition in fission yeast. *Nature* **305**: 682.

Egel, R. 1984. The pedigree pattern of mating-type switching in *Schizosaccharomyces pombe*. *Curr. Genet.* **8**: 205.

Egel, R. and H. Gutz. 1981. Gene activation by copy transposition in mating-type switching of a homothallic fission yeast. *Curr. Genet.* **3**: 5.

Egel, R., D.H. Beach, and A.J.S. Klar. 1984. Genes required for initiation and resolution steps of mating-type switching in fission yeast. *Proc. Natl. Acad. Sci.* **81**: 3481.

Miyata, H. and M. Miyata. 1981. Mode of conjugation in homothallic cells of *Schizosaccharomyces pombe*. *J. Gen. Appl. Microbiol.* **27**: 365.

Shibata, T., H. Watabe, T. Kaneko, I. Iino, and T. Ando. 1984. On the nucleotide sequence recognized by a eukaryotic site-specific endonuclease, Endo.*Sce*I from yeast. *J. Biol. Chem.* **259**: 10499.

Initiation and Resolution Steps of Recombination for Yeast Mating-type Inconversion

A.J.S. Klar
Cold Spring Harbor Laboratory
Cold Spring Harbor, New York 11724

Studies of genetic recombination in eukaryotes have been conducted primarily on fungi in which all four products of each meiosis can be obtained and analyzed. The organisms of choice have been *Ascobulus immersus, Saccharomyces cerevisiae* (budding yeast), and *Schizosaccharomyces pombe* (fission yeast). To complement classical genetic approaches, recent interest has focused on defining the biochemical parameters, which include initiation as well as the resolution steps of recombination. Over the years, we have used the mating-type switching systems of budding yeast and fission yeast as the model systems for genetic recombination. The high efficiency with which both systems recombine has allowed us to define precisely the nature of the initiation event, and comparison of these evolutionarily quite unrelated systems may shed light on the differences in the resolution steps.

The *S. cerevisiae MAT* Interconversion

The alternate alleles of the mating-type locus, called *MAT*a and *MAT*α, interconvert by a transposition-substitution event where a copy of either the *HML* or *HMR* donor information is transmitted to the expressed *MAT* locus (for reviews, see Herskowitz and Oshima 1981; Haber 1983; Nasmyth 1983a; Klar et al. 1984b). Three points are worth noting regarding the discussion of initiation and the resolution steps of recombination. First, the initiation event for recombination constitutes a double-chain break found at *MAT* at a specific sequence (Strathern et al. 1982; Kostriken et al. 1983; see also Malone and Esposito 1980; Weiffenbach and Haber 1981). The unlinked *HO* gene has been shown to code for the so-called Y/Z endonuclease, which catalyzes the break in vivo (Kostriken and Heffron 1984). Furthermore, the break cannot be repaired in the absence of the donor loci. Such strains produce inviable cells in a predictable fashion; that is, only those cells that attempt to switch die, thus generating the so-called pedigree of death (Klar et al. 1984a). Second, the pattern of switching in a cell lineage is very precise. Only the mother cells divide to generate two switched cells in 86% of cell divisions; the daughter (bud) cells are incompetent for switching (Hicks and Herskowitz 1976; Strathern and Herskowitz 1979). This pattern of switching implies that both chromatids produced by the mother cell inherit newly switched and identical *MAT* alleles. Third, the resolution of the gene conversion event required for switching is constrained such that recombination of flanking markers is not allowed in this intrachromosomal event. (The *HML* and *HMR* donor loci are located on the same chromosome where *MAT* resides [Harashima and Oshima 1976].) All these features are observed in cells undergoing switching during mitotic growth.

In contrast, if *MAT* conversion involved homologs, we frequently found recombination of flanking markers associated with the gene conversion event. When a diploid of genotype *hmlΔ/hmlΔ MATα-inc/mata$^-$ hmrΔ/hmrΔ HO/HO* was allowed to change to *MATα-inc/MATα-inc* during mitotic growth, nearly 25% of the events were associated with recombination of flanking markers (Klar and Strathern 1984). (The symbol Δ indicates deletion and *MATα-inc* is an allele that fails to get cleaved in vivo and thus cannot switch.) Our similar recent experiments, but in cells undergoing meiosis, also demonstrate that gene conversion events are associated with recombination of flanking markers (A.L. Kolodkin et al., in prep.). We find that when the *HO* gene function was provided to diploid cells (genotype *hmlΔ/hmlΔ MATα-inc/MATα hmrΔ/hmrΔ*) undergoing meiosis, as many as 14% of the tetrads generated four *MATα-inc*:0 *MATα* conversions. Interestingly, no 3:1 events were obtained. Nearly 30% of the time the closely linked marker *tsm1* was also found to coconvert.

The *S. pombe mat1* Interconversion

The two mating types, designated P (plus) and M (minus), are shown to interconvert by a transposition-substitution event where a copy of either the *mat2-P* or the *mat3-M* cassette is transmitted to the expressed *mat1* locus (see description of the system by R. Egel, this volume). Although the *S. pombe* system is formally analogous to that found in *S. cerevisiae*, specific details are quite different and promise to provide information regarding the evolution of both systems. Some of those differences are highlighted below.

Like the *S. cerevisiae* system, the *S. pombe mat1* switching is initiated by a double-stranded DNA break found at *mat1* (Beach 1983). Recently, we constructed strains deleted for both donor loci. The donor-deleted strains have a wild-type level of the break, yet do not produce inviable cells (A.J.S. Klar and L.M. Miglio, in prep.). Therefore, the broken ends at *mat1* can be repaired (healed) by this yeast even without *mat1* switching.

The pattern of switching in an *S. pombe* cell lineage is quite different from that found in *S. cerevisiae*. Miyata and Miyata (1981) found that in *S. pombe* only one cell among four progeny of a single cell (obtained after growth of two generations) is competent to switch. Recent work has shown that this asymmetry arises because of some semiheritable change(s) in the chromosome at *mat1* (Egel 1984a and this volume; A. Klar, unpubl.). In contrast, the precise pattern in *S. cerevisiae* is dictated by the *HO* gene; it is expressed in mother cells but not in daughter cells (Nasmyth 1983b).

Like the *S. cerevisiae* system, *mat1* conversion in *S. pombe* is an intrachromosomal transposition event and it is found to occur without the recombination of flanking markers. (The *mat2* and *mat3* donor loci are closely linked to *mat1* [Egel 1984b].) However, when we meiotically crossed strains of the opposite mating type which are deleted for the donor loci, as many as 20% of the asci generated 3:1 + 1:3 conversions at *mat1*. These conversion events were frequently associated with the recombination of flanking markers. Strains that did not exhibit the double-stranded break at *mat1* failed to gene-convert (A.J.S. Klar and L.M. Miglio, in prep.). Thus, as in the *S. cerevisiae MAT* switching system, the double-stranded break in *S. pombe* initiates recombination efficiently both in meiosis as well as in mitosis.

However, major differences, possibly in the resolution steps of meiotic gene conversion at the mating-type locus in both yeasts, are indicated by the difference in the type of converison events obtained. In *S. cerevisiae* only 4:0 types are obtained, whereas only 3:1 + 1:3 are observed in *S. pombe*. The reason for these differences is not known, but these results are consistent with the features of the switching pattern observed in mitotically dividing cells. In *S. cerevisiae*, switches occur in pairs of cells (Hicks and Herskowitz 1976; Strathern and Herskowitz 1979); therefore both chromatids of a parent cell must have acquired the potential to switch. If such a control is also operative in meiosis, it will dictate the observed 4:0 pattern. In the *S. pombe* system, among a pair of sister cells, only one member generates a single switched cell in the subsequent generation (Miyata and Miyata 1981). In other words, among a pair of sister chromatids, only one member switches. If such a constraint is also applied to meiotic conversions at *mat1*, only the observed 3:1 and 1:3 events would be expected. We imagine that the presence of only 3:1 and 1:3 meiotic conversion classes in *S. pombe* and only 4:0 in *S. cerevisiae* reflect major differences in molecular details of the systems, possibly due to differences in the time of generating breaks, repair, and the resolution step of recombination. It should be noted that the gene conversion events found at many other loci in *S. cerevisiae* are primarily of 3:1 and 1:3 type and not of 4:0 and 0:4 type (Fogel et al. 1981). To explain the differences in the behavior of *MAT* conversion with other loci, we have to entertain the possibility that conversions at other loci are not initiated by the double-stranded breaks in *S. cerevisiae*. Further studies are clearly needed to resolve this issue.

It is interesting to compare the time of generation, time of utilization of the break, and the possibility of protection of broken ends in both systems. In *S. cerevisiae*, in exponentially growing culture only about 2% of *MAT* DNA is found to be cleaved (Strathern et al. 1982), because it is healed quickly with *MAT* switching. In *S. pombe*, as much as 20% of DNA is found to be broken in all stages of the cell cycle (Beach 1983). The available data suggest that the break is made in the S period and that it persists throughout the whole length of the cell cycle. The break is repaired by *mat1* switching in the next S period. It is likely that the cut ends in *S. pombe* are protected from nucleolytic degradation. Even strains that are deleted for the donor loci have the break, which must be repaired independent of switching. Possibly the activity that catalyzes the break may protect the broken ends by holding them together. The same activity may help ligate the ends, even without *mat1* switching. Such a protection and/or ligation mechanism may not be operative in *S. cerevisiae* because strains deleted for the donor loci are known to produce inviable cells (Klar et al. 1984b). In this system, the healing may occur only by a successful switching event.

In summary, both yeasts efficiently accomplish the gene conversion event, which is shown to be initiated by a double-stranded break. However, all the details of the mechanism of switching are quite different and their comparison has already produced a wealth of information for elucidating the mechanism of recombination.

ACKNOWLEDGMENT

This work was supported by a grant (GM 25678) from the National Institutes of Health.

REFERENCES

Beach, D.H. 1983. Cell type switching by DNA transposition in fission yeast. *Nature* **305:** 682.

Egel, R. 1984a. The pedigree pattern of mating-type switching in *Schizosaccharomyces pombe*. *Curr. Genet.* **8:** 205.

―――. 1984b. Two tightly linked silent cassettes in the mating-type region of *Schizosaccharomyces pombe*. *Curr. Genet.* **8:** 199.

Fogel, S., Mortimer, R.K., and K. Lusnak. 1981. Mechanisms of meiotic gene conversion or "wanderings on a foreign strand." In *The molecular biology of the yeast* Saccharomyces: *Life cycle and inheritance* (ed. J.N. Strathern et al.), p. 289. Cold Spring Harbor Laboratory, Cold Spring Harbor, New York.

Haber, J.E. 1983. Mating-type genes of *Saccharomyces cerevisiae*. In *Mobile genetic elements* (ed. J. Shapiro), p. 559. Academic Press, New York.

Harashima, S. and Y. Oshima. 1976. Mapping of the homothallism genes *HMa* and *HMα* in *Saccharomyces* yeast. *Genetics* **84:** 437.

Herskowitz, I. and Y. Oshima. 1981. Control of cell types in *Saccharomyces cerevisiae*: Mating-type and mating-type interconversion. In *The molecular biology of the yeast* Saccharomyces: *Life cycle and inheritance* (ed. J.N. Strathern et al.), p. 181. Cold Spring Harbor Laboratory, Cold Spring Harbor, New York.

Hicks, J.B. and I. Herskowitz. 1976. Interconversion of yeast mating types. Direct observation of the action of the homothallism (*HO*) gene. *Genetics* **83:** 245.

Klar, A.J.S. and J.N. Strathern. 1984. Resolution of recombination intermediates generated during yeast mating-type switching. *Nature* **310:** 744.

Klar, A.J.S., J.N. Strathern, and J.A. Abraham. 1984a. The involvement of double-strand chromosomal breaks for mating-type switching in *Saccharomyces cerevisiae*. *Cold Spring Harbor Symp. Quant. Biol.* **49:** 77.

Klar, A.J.S., J.N. Strathern, and J.B. Hicks. 1984b. Developmental pathways in yeast. In *Microbial development* (ed. R. Losick and L. Shapiro), p. 151. Cold Spring Harbor Laboratory, Cold Spring Harbor, New York.

Kostriken, R. and F. Heffron. 1984. The product of the *HO* gene is a nuclease: Purification and characterization of the enzyme. *Cold Spring Harbor Symp. Quant. Biol.* **49:** 89.

Kostriken, R., J.N. Strathern, A.J.S. Klar, J.B. Hicks, and F. Heffron. 1983. A site specific endonuclease essential for mating-type switching in *Saccharomyces cerevisiae*. *Cell* **35:** 167.

Malone, R.E. and R.E. Esposito. 1980. The *RAD52* gene is required for homothallic interconversion of mating types and spontaneous mitotic recombination in yeast. *Proc. Natl. Acad. Sci.* **77:** 503.

Miyata, H. and M. Miyata. 1981. Modes of conjugation in homothallic cells of *Schizosaccharomyces pombe*. *J. Gen. Appl. Microbiol.* **27:** 365.

Nasmyth, K.A. 1983a. Molecular genetics of yeast mating-type. *Annu. Rev. Genet.* **16:** 439.

―――. 1983b. Molecular analysis of cell lineage. *Nature* **302:** 670.

Strathern, J.N. and I. Herskowitz. 1979. Asymmetry and directionality in production of new cell types during clonal growth: The switching pattern of homothallic yeast. *Cell* **14:** 372.

Strathern, J.N., A.J.S. Klar, J.B. Hicks, J.A. Abraham, J.M. Ivy, K.A. Nasmyth, and C. McGill. 1982. Homothallic switching of yeast mating type cassettes is initiated by a double-stranded cut in the *MAT* locus. *Cell* **31:** 183.

Weiffenbach, B. and J.E. Haber. 1981. Homothallic mating type switching generates lethal chromosomal breaks in *rad52* strains of *Saccharomyces cerevisiae*. *Mol. Cell. Biol.* **1:** 522.

Intermediates in Homothallic Switching

J. Strathern, C. McGill, B. Shafer, and D. Raveh

NCI-Frederick Cancer Research Facility
LBI-Basic Research Program, Frederick, Maryland 21701

The *HO* endonuclease initiates homothallic switching in *Saccharomyces cerevisiae* by making a double-stranded cut in the *MAT* locus (Strathern et al. 1982). Homothallic switching from the *MAT*a to the *MAT*α allele (or *MAT*α to *MAT*a) has several aspects analogous to gene conversion. However, this process has unique features that facilitate genetic and biochemical analyses. These include its high efficiency (about 40% of the cells in each division), the clear definition of the donor (the unexpressed *HML* or *HMR* copies of the *MAT* DNAs) and the recipient (*MAT*), and, finally, the fact that the *MAT*, *HML*, and *HMR* loci are cloned and sequenced as substrates. Our studies are based on the philosophy that the detailed analysis of the physical and genetic consequences of this process at a unique site can help us understand how such double-strand-break-initiated events might contribute to generalized meiotic and mitotic gene conversion and recombination.

Our recent work deals with the determination of the extent of the DNA at *MAT* that is changed during the switching process. The original genetic observations and physical analysis demonstrated that the *MAT*α allele differed from the *MAT*a allele because of DNA sequence substitutions (Fig. 1). The 747-bp *MAT*α-specific sequence (Yα) is replaced by a 642-bp Ya-specific sequence from a donor (usually *HMR*a) each time an α-to-a switch occurs. Flanking the allele-specific Y regions are regions X and Z found at both donors. Because these sequences are identical at *MAT*, *HML*, and *HMR*, it was not possible to determine how much more than the Y region was switched or whether the amount was variable. Some of the results reviewed here were obtained by using restriction site polymorphisms in the X and Z regions as markers to determine the consequences of switching on the DNA in these regions.

Physical and genetic studies demonstrated that homothallic switching occurs after G_1 and before the replication of the *MAT* locus. The switch is followed by DNA replication so that two switched cells are produced. Because there is a one-generation de-

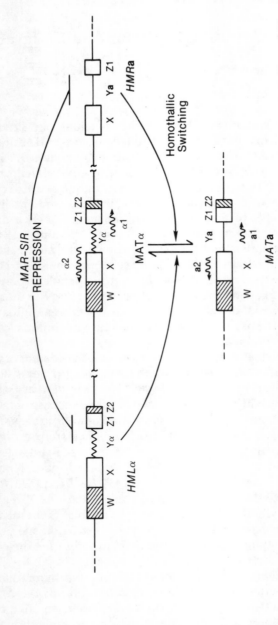

Figure 1 Diagram of the mating-type cassettes on chromosome III. Diagram is not to scale. W, X, Y, Z1, and Z2 represent regions defined by homologies between *MAT* and *HML* or *HMR*. The W region is 723 bases long and is homologous between *HML* and *MAT*. The X region (704 bp) is found at *HML*, *MAT*, and *HMR*. *HML*, *MAT*, and *HMR* can have either the **a**-specific sequence Y**a** (642 bp) or the α-specific sequence Yα (747 bp). Z1 (239 bp) is found at *HML*, *MAT*, and *HMR*, whereas Z2 (89 bp) is found only at *HML* and *MAT*.

lay for spores or fresh buds before the *HO* gene can be expressed (Hicks and Herskowitz 1976), the switches are produced in a very ordered pattern in clonal pedigrees (Strathern and Herskowitz 1979). One additional feature of the homothallic switching system that facilitated this analysis is that the *HO* gene is regulated by cell type: *HO* is expressed in **a** and α cells but not in **a**/α diploids (Jensen et al. 1983).

We used the switching pedigrees and the cell-type regulation of the *HO* gene to freeze the results after single homothallic switches. The starting strains were homothallic α spores that had different restriction sites in the X region between *MAT*α and *HMR***a**. These spores were allowed to grow to the four-cell stage, at which time many had undergone one switch to produce two *MAT***a** cells. These cells mated with their *MAT*α siblings to produce two zygotes. The zygotes were micromanipulated apart and allowed to grow into colonies. Because the *HO* gene is turned off in these **a**/α zygotes, no more switches occur. The clones devised from such zygotes can be grown into cultures and the DNA can be isolated and analyzed. The two zygotes represent the two DNA strands at *MAT* after the switch and before DNA replication. Hence, as shown below, we were able to detect heteroduplexes made during switching at the site of the restriction site polymorphisms.

We monitored the consequences of switching on four separate sites in the X region and one site in the Z region. Although the results can be summarized fairly succinctly, we believe they significantly define the mechanism of homothallic switching and eliminate the extremes of some models.

1. Sequences in X and Z can be coconverted with the *MAT*α to *MAT***a** switch (*HMR***a** as donor).
2. Sites close to the Y region have a higher probability of coconversion than sites farther away.
3. In the pairs of zygotes from single switches, examples of two coconversions of the restriction site with *MAT* (two-strand change), one coconversion (heteroduplex formation), and no coconversion are found.
4. The sequences from *MAT* were not transferred to *HMR*.
5. No recombinations between *MAT* and *HMR* were detected.

We draw several conclusions from these data. First, the extent of the coconversion and hence the site of resolution is variable, with the probability falling off as a function of distance. This observation is reminiscent of meiotic gradients of gene conversion (Fogel et al. 1981). Second, the absence of transfer of the *MAT* restriction site markers to *HMRa* suggests that no stable reciprocal heteroduplex DNA is made. In this sense homothallic switching further mimics meiotic conversion in yeast. Third, the absence of recombination between donor and recipient suggests that some mechanism constrains the resolution of the switching process.

We suggested that the 3' strand from the Z region side of the *HO* cut melts into *HMRa* and serves as the primer for synthesis of the new copy. Of particular interest in this regard are the observations that coconversion of sequences in Z could occur and that Z-region polymorphisms could not be transferred to *HMR*. The restriction site marker that we made for this experiment was close to the *HO* cut (a single-base change 23 bases from the 3' end) in the hope of seeing evidence of the "primer" left behind on *HMR*. The failure to see this transfer can have several different explanations: (1) The primer could always be on the duplex inherited by *MAT*; (2) any mismatched bases left at *HMR* could be corrected in the direction of the donor; (3) the primer (and its new copy of Ya) could be pulled out of *HMR*, restoring the original duplex; (4) the resolution site could be fewer than 23 bases from the cut; and (5) the primer could be degraded past the marker base.

The simplest form of the double-strand-break model has as an intermediate a DNA bubble involving two Holliday structures and strand exchanges between donor and recipient (Szostak et al. 1983). This intermediate has twofold symmetry; that is, if there is a heteroduplex on the left side on the recipient, there is a heteroduplex on the right side on the donor.

Because we find evidence of heteroduplex formation on both sides of the Y region of *MAT*, but on neither side of *HMR*, we conclude that homothallic switching does not occur by cutting and religating this kind of intermediate. The results we obtained can be derived from this intermediate if it is resolved by replication. That mechanism would reform the original duplex at *HMR* and place all of the newly synthesized DNA at *MAT*. Heteroduplex DNA would be the result of differential chewback and resynthesis of the 5' and/or 3' DNAs on both sides of the Y region. The replicative resolution of the bubble would result in those heteroduplexes ending up at *MAT*. This mechanism has the further property that it would occur without recombination of the donor with the recipient.

The genetic observations presented above suggest that there is loss of sequences in the Z region during switching. This could occur by chewback of the DNA from the site of the original *HO* cut. The role of the 3' end as a primer could be facilitated by chewback of the 5' strand. For these reasons, we are developing techniques to monitor the ends of the DNA produced by the *HO* cut in vivo. We have two goals: first, to detect the *HO* cut to one-base resolution in vivo so that we can observe degradation or elongation from that site; and second, to divorce homothallic switching from its normal association with the S phase of the cell cycle.

Our previous detection of the *HO* cut in vivo was by the Southern blot technology. This gave us resolution of about 20 bases but could not detect whether the site of the cut was stable (Strathern et al. 1982). We have been able to do primer extensions on total yeast genomic DNA and detect the *HO* cut (Fig. 2). This technology de-

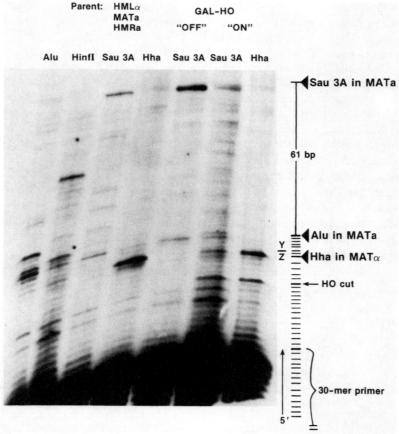

Figure 2 Total genomic yeast DNA was cleaved with the indicated enzyme, denatured, and hybridized to an end-labeled synthetic oligonucleotide primer. The primer was extended with Klenow fragment of DNA *pol*I and analyzed on a polyacrylamide gel. The two rightmost tracks are from G_0 cells in which the *HO* gene is induced.

tects the steady-state level of the 5' ends of the cut DNA. These results confirm the position of the cut defined by in vitro analysis as being between the seventh and eighth bases of the Z region.

Further experiments will be required to determine whether the subsequent processing of this cut can be detected. Normally the *HO* cut occurs at the point in the cell cycle at which cells are committed to DNA replication (Nasmyth 1983). We would like to examine the DNA synthesis that is associated specifically with the *MAT* cassette gene conversion process. Our first attempts used the *HO* gene under the control of the *GAL* promoter (courtesy of R. Jensen). These experiments demonstrated that the *GAL* promoter can be induced in G_0-arrested cells produced by growth to saturation or in G_1-arrested cells produced by the use of α-factor. For example, the *HO* "ON" DNA in Figure 2 is from G_0 cells. Our observations so far suggest that G_0 cells are not capable of completing a homothallic switch despite the fact that up to 50% of the *MAT* DNA becomes cleaved in vivo by *HO*. We can detect no loss of *MAT* sequences in this experiment, suggesting that the ends are stable. When these cells are returned to growth conditions, efficient switching and good viability result. Analogous experiments with GAL:HO *MAT*a cells arrested with α-factor are in progress.

It is our expectation that the combined approaches of fine-structure genetic analysis and physical monitoring of the homothallic switching process will allow us to define the pathways by which it occurs. It is our hope that we can combine that information with mutants defective in homothallic switching and double-strand-break repair to determine the roles of these functions in the recombination processes.

ACKNOWLEDGMENTS

This research was sponsored by the National Cancer Institute, DHHS, under contract No. NO1-CO-23909 with Litton Bionetics, Inc. The contents of this publication do not necessarily reflect the views or policies of the Department of Health and Human Services, nor does mention of trade names, commercial products, or organizations imply endorsement by the U.S. Governmnent.

REFERENCES

Fogel, S., R.K. Mortimer, and L. Lusnak. 1981. Mechanisms of meiotic gene conversion, or "wanderings on a foreign strand." In *The molecular biology of the yeast* Saccharomyces: *Life cycle and inheritance* (ed. J.N. Strathern et al.), p. 289. Cold Spring Harbor Laboratory, Cold Spring Harbor, New York.

Hicks, J. and I. Herskowitz. 1976. Interconversion of yeast mating types. I. Direct observation of the action of the homothallism (*HO*) gene. *Genetics* **83:** 245.

Jensen, R., G.F. Sprague, Jr., and I. Herskowitz. 1983. Regulation of yeast mating-type interconversion: Feedback control of *HO* gene expression by the mating-type locus. *Proc. Natl. Acad. Sci.* **80:** 3035.

Nasmyth, K. 1983. Molecular analysis of a cell lineage. *Nature* **302:** 670.

Strathern, J.N. and I. Herskowitz. 1979. Asymmetry and directionality in production of new cell types during clonal growth: The switching pattern of homothallic yeast. *Cell* **17:** 371.

Strathern, J.N., A.J.S. Klar, J.B. Hicks, J.A. Abraham, J.M. Ivy, K.A. Nasmyth, and C. McGill. 1982. Homothallic switching of yeast mating-type cassettes is initiated by double-stranded cut in the *MAT* locus. *Cell* **31:** 183.

Szostak, J.W., T.L. Orr-Weaver, and R.J. Rothstein. 1983. The double-strand-break repair model for recombination. *Cell* **33:** 25.

Stimulation of Mitotic Recombination by *HO* Nuclease

J.A. Nickoloff, E. Chen, and F. Heffron
Scripps Clinic and Research Foundation
La Jolla, California 92037

HO nuclease is a site-specific, double-strand endonuclease present in haploid cells of *Saccharomyces cerevisiae* undergoing mating-type interconversion (Hicks et al. 1977; Herskowitz and Oshima 1981; Jensen and Herskowitz 1984). *HO* initiates mating-type interconversion by making a double-strand break within the *MAT* locus leaving a 4-bp 3' extension at a site in a region homologous to *HML* and *HMR* (Z) near the Y/Z junction (see Fig. 1). The cleavage site in vitro corresponds to the site in vivo (Kostriken et al. 1983; Kostriken and Heffron 1984). Mutations in *MAT* that eliminate mating-type interconversion are found to be close to the cut site and are more refractory to in vitro digestion (Kostriken et al. 1983; Weiffenbach et al. 1983). The sensitivity of various deleted substrates to the *HO* endonuclease had suggested that the enzyme recognized, at most, a 35-bp sequence immediately adjacent to the Y/Z junction in *MAT* (Strathern et al. 1982; Kostriken et al. 1983; Weiffenbach et al. 1983; Kostriken and Heffron 1984). To define precisely the recognition site for *HO* we have constructed a number of mutations and deletions within or adjacent to the site. The minimal recognition site is 18 bp. However, several shorter substrates and many substrates containing point mutations are cleaved at low, but detectable, levels in vitro. In this study we found the left-hand 6 bp of the *HO* recognition site to be present in *MAT*a and to lie within a region of *MAT* that is different in **a** and α cells.

We are interested in testing the hypothesis that high efficiency of mating-type switching is due solely to the efficient generation of a double-strand break at *MAT*. A copy of a 24-bp sequence recognized by *HO* in vitro stimulates mitotic recombination between *ura3* heteroalleles in an *HO* haploid strain of yeast, but unlike mating-type interconversion, the recombination between the *ura3* heteroalleles takes place with exchange of flanking markers.

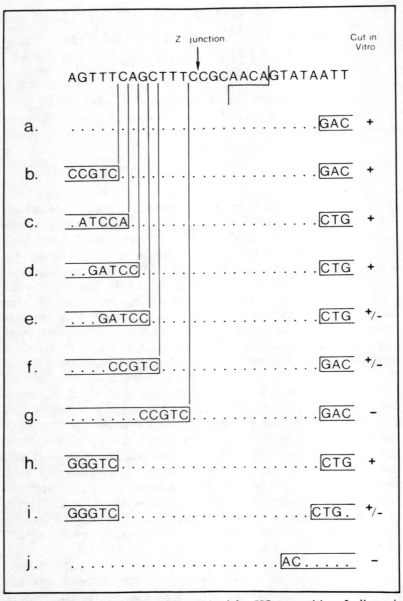

Figure 1 Sequences in Y and Z required for *HO* recognition. In line a is shown a sequence that contains 67 bp of Ya and 12 bp of Z. The dots indicate identity with the sequence at the top, and the boxed regions indicate vector sequences. On the right are shown the results of in vitro digestion by *HO*.

RESULTS

To test the susceptibility of various substrates to *HO* digestion, we have used *HO* nuclease partially purified by column chromatography. The *Escherichia coli* plasmid encoding *HO* (pRK128) produces a protein identical to that made in *S. cerevisiae* (Kostriken and Heffron 1984). We have tested each of the substrates shown in Figure 1 for digestion by *HO* nuclease in vitro. The deletions in Y and Z were constructed in one of two ways. The substrates shown in b, f, g, and j of this figure were constructed by BAL-31 digestion of *MATa*. The other substrates were constructed by inserting synthetic oligonucleotides in pUC19.

To assay for *HO* digestion in vitro, we used a unique restriction site to linearize plasmids and then treated the DNA with *HO* nuclease. The sizes of the resulting fragments were diagnostic for DNA cut at the correct site. In the following discussion we have numbered the bases in the allele-specific DNA (Y) and the sequences to the right of the Y/Z junction (Z) starting with the Y/Z junction (e.g., Y1, Z6). The substrate shown in line a of Figure 1 contains 12 bp of Z and 66 bp of Y. It is digested by *HO* to the same degree as an intact clone of *MATa*. A substrate leaving bases Y1–Y8 intact reduces in vitro digestion by *HO* endonuclease approximately twofold (line b). Transversion mutations at either base Y7 or Y8 are digested at most fourfold less than a complete copy of *MAT* (lines c and d).

Mutations at base Y6 have a more pronounced effect on in vitro digestion. The substrates shown in lines e and f are identical for base pairs Y1–Y5 but contain mutations at base Y6. The substrate in line e contains a G→C transversion and is digested at least 10-fold less than an intact substrate. The substrate in line f contains a G→T transversion at the same base and also shows reduced digestion. A substrate containing only Y1 and Z1–Z12 was not appreciably digested by the endonuclease in vitro (line g). Thus, Y6 appears to be the first base in Y that has a pronounced effect on in vitro digestion by *HO*.

The effects of changes in Z on in vitro digestion by *HO* were tested using substrates shown in lines h, i, and j. The substrate shown in h is digested approximately twofold less than a complete copy of *MATa*. Plasmid DNA containing the sequence shown in line i, however, is digested at least 10-fold less than a complete copy of *MATa*. Substitution of a C at position Z11 behaves similarly to the C substitution at position Z12 and is digested about 10-fold less well than an intact substrate. A substrate that retains only 7 bp of Z is completely refractory to digestion by the enzyme (line j). A sequence polymorphism at base Z11 had been noted in comparing the sequences of several clones of mating-type sequence (Astell et al. 1981). The polymorphism has an A at Z11 instead of a T. This change makes the substrate more refractory to digestion in short substrates such as those shown, but has no effect when sequence homology to *MAT*a is extended by another 4 bp. Thus, for the *MAT*a allele shown, at least 12 bp of Z are required for digestion by *HO* nuclease in vitro.

We compared the digestion of intact *MAT*a and *MAT*α substrates. Both appear to be equally good substrates for the enzyme. The digestion of two cloned oligonucleotides that correspond to either *MAT*a or *MAT*α for bases Y8–Z12 were compared. The *MAT*α substrate is digested at least fivefold less completely than the *MAT*a substrate. This result suggests that *HO* nuclease requires a longer *MAT*α recognition sequence than a *MAT*a sequence.

*MAT*α and *MAT*a show four base-pair differences within the first eight bases of Y. We constructed a hybrid *MAT* substrate containing a mixture of **a** and α allelic bases. This substrate is digested by *HO* at a level intermediate between that of either parent, indicating that the enzyme does not simply recognize a set of four allelic bases from either **a** or α. A *MAT*a substrate that contains a transversion at base Y3 is a poor substrate. Similarly, a substrate containing a transversion at base Y2 is also a poor substrate. These results should be considered in light of the earlier results for sequence requirements of Z. Both of these results suggest that the farther away a mutation is located from the Y/Z boundary, the less effect it has on digestion by *HO*.

We inserted a synthetic *HO* site within one of two *ura3* heteroalleles to test whether a site cleaved by *HO* in vitro will stimulate recombination in vivo. The structures of the *ura3* heteroalleles are shown in Figure 2. These strains are deleted for *MAT* and contain a plasmid encoding a galactose-inducible *HO* gene. The strains were grown to log phase in medium containing glycerol as a carbon source. Equal aliquots were diluted 1:2 and grown for 6 hours in fresh medium containing either glucose or galactose as a carbon source to provide repression or induction of *HO* expression, respectively. Cells were harvested, washed, and plated on uracil-omission and -rich (YPD) plates. Ura^+ colonies were counted and tested for their *His* phenotype. The results are given in Table 1. The 24-bp sequence stimulates recombination 80- to 100-fold over spontaneous levels. No stimulation is observed when the *HO* site is replaced with an *Eco*RI linker mutation alone (JY13).

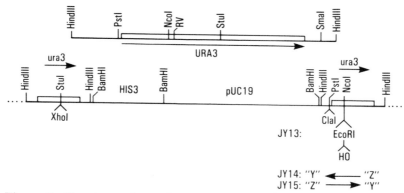

Figure 2 Structures of *ura3* heteroalleles. (*Top*) Structure of *URA3* gene region. (*Bottom*) Structures of *ura3* heteroalleles. All strains have an *Xho*I linker insertion mutation in the *Stu*I site of one gene, and a *Cla*I linker insertion mutation in the *Pst*I site upstream of the second gene. The *Cla*I linker introduces an out-of-frame ATG start codon. JY13 contains an *Eco*RI linker insertion mutation in the *Nco*I site of the second gene. JY14 and JY15 contain 24-bp *HO* recognition sites inserted in the *Eco*RI site of JY13. The orientation of the *HO* sites for these strains is shown at the bottom.

JY14 and JY15 have the *HO* site in opposite orientations. The recognition site itself is not palindromic and one might predict that the orientation has an effect on recombination. We observe no difference in recombination frequency depending on the *HO* site orientation. Approximately 10% of the Ura^+ recombinants are His^+ when grown in glucose and 20% are His^+ when induced. We have no explanation for the increase in number of His^+ recombinants following induction.

DISCUSSION

We have determined that the recognition site for *HO* nuclease includes bases that are 18 bp apart. This includes bases within the Y,

Table 1 Effect of *HO* Nuclease on Mitotic Recombination

Strain	Recombination frequency		His^+/Ura^+	
	glucose	galactose	glucose	galactose
JY13	4.0×10^{-4}	4.0×10^{-4}	nt[a]	nt
JY14	5.8×10^{-4}	4.8×10^{-2}	0.23 (292)	0.37 (239)
JY15	5.7×10^{-4}	4.2×10^{-2}	0.14 (287)	0.23 (208)

Recombination frequency is expressed as the number of Ura^+ colonies per cell plated. His^+/Ura^+ is the ratio of His^+ to Ura^+ colonies. The numbers in parentheses indicate the number of colonies tested.
[a]nt, Not tested.

or allele-specific, region of *MAT*. Most of the single base-pair changes we have tested show reduced, but appreciable, digestion by the enzyme. This suggests that in vitro the enzyme does not have a precise recognition site, unlike prokaryotic type-II restriction enzymes. The recognition of *HO* sites by *HO* nuclease may be more closely related to other DNA binding proteins, such as *CAP*, and λ repressor, which recognize a consensus sequence and must make contact at a certain number of bases before a reaction can take place. A 24-bp recognition site stimulates recombination when it is introduced into an *HO* haploid strain of yeast. A single base-pair change at *MAT* can make the strain inconvertible and relieves the lethality of *MAT*α in a *rad52 HO* background (Malone and Esposito 1980; Weiffenbach and Haber 1981). This suggests that no other recognition sites are present within the yeast genome. In agreement with this, *HO* has not been observed to stimulate recombination at any sites in vivo, other than *MAT*. Perhaps sequences that are cut weakly by *HO* have been eliminated from the yeast genome. Alternatively, the enzyme may have a lower specificity in vitro than it has in vivo.

A *MAT***a** cell preferentially replaces the allele-specific sequences from *MAT* with those present at the *HML*α cassette (Klar et al. 1982). The greater number of bases of homology required for *HO* recognition of *MAT*α than *MAT***a** suggests that *HO* must be capable of discriminating between the two alleles. It is possible that *HO* plays a role in determining the directionality of switching.

We find that an *HO* recognition site, identified by in vitro studies, can stimulate recombination between two *ura3* heteroalleles approximately two orders of magnitude. This stimulation is much lower than the frequency of mating-type interconversion. There are at least three possible explanations for this: (1) *HO* may be poorly induced in this strain because of a defect in galactose induction; (2) the site that we used may not be digested by *HO* as well as one that is slightly larger in vivo because it is missing contacts, either linked or unlinked, to the site we have introduced; (3) the recognition site at *MAT*, but not the short sequence we have used, is a DNA-hypersensitive site that is controlled in some way by the products of *MAR* and *SIR*.

During mating-type interconversion, a double-strand break is introduced at *MAT* in the Z region immediately adjacent to the allele-specific sequences. The recognition site for *HO* is not symmetric and interconversion involves asymmetry (cf. sequences in Y must be removed but sequences in Z cannot be removed). However, we do not find a difference in recombination frequency depending on orientation. Thus, recombination is most likely stimulated only because a double-strand break has been made.

Ura$^+$ recombinants were isolated that had arisen both by crossing-over and by gene conversion. The recombination that we observe here is therefore identical to that which has been observed for recombination stimulated by transformed linear DNA. The recombination, however, is different from that observed during mating-type interconversion. We hypothesize that there may be additional sequences or *trans*-acting factors which determine that mating-type interconversion takes place by a mechanism without exchange of flanking markers.

ACKNOWLEDGMENTS

We thank Deborah Beutler for technical assistance. This work was funded by Public Health Service grant RO1 GM33808-01 from the National Institutes of Health.

REFERENCES

Astell, C.R., L. Ahlstrom-Jonasson, M. Smith, K. Tatchell, K.A. Nasmyth, and B.D. Hall. 1981. The sequence of the DNAs coding for the mating type loci of *Saccharomyces cerevisiae*. *Cell* **27**: 15.

Herskowitz, I and Y. Oshima. 1981. Control of cell type in *Saccharomyces cerevisiae*: Mating type and mating-type interconversion. In *The molecular biology of the yeast* Saccharomyces: *Life cycle and inheritance* (ed. J.N. Strathern et al.), p. 181. Cold Spring Harbor Laboratory, Cold Spring Harbor, New York.

Hicks, J.B., J.N. Strathern, and I. Herskowitz. 1977. The cassette model of mating-type interconversion. In *DNA insertion elements, plasmids, and episomes* (ed. A.I. Bukhari et al.), p. 457. Cold Spring Harbor Laboratory, Cold Spring Harbor, New York.

Jensen, R.E. and I. Herskowitz. 1984. Directionality and regulation of cassette substitution in yeast. *Cold Spring Harbor Symp. Quant. Biol.* **49**: 97.

Klar, A.J.S., J.B. Hicks, and J.N. Strathern. 1982. Directionality of yeast mating-type interconversion. *Cell* **28**: 551.

Kostriken, R. and F. Heffron. 1984. The product of the HO gene is a nuclease: Purification and characterization of the enzyme. *Cold Spring Harbor Symp. Quant. Biol.* **49**: 89.

Kostriken, R., J.N. Strathern, A.J.S. Klar, J.B. Hicks, and F. Heffron. 1983. A site-specific endonuclease essential for mating-type switching in *Saccharomyces cerevisiae*. *Cell* **35**: 167.

Malone, R. and R.E. Esposito. 1980. The *RAD52* gene is required for homothallic interconversion of mating types and spontaneous mitotic recombination in yeast. *Proc. Natl. Acad. Sci.* **77**: 503.

Strathern, J.N., A.J.S. Klar, J.B. Hicks, J.A. Abraham, J.M. Ivy, K.A. Nasmyth, and C. McGill. 1982. Homothallic switching of yeast mating type cassettes is initiated by a double stranded cut in the *MAT* locus. *Cell* **31**: 183.

Weiffenbach, B. and J.E. Haber. 1981. Homothallic mating type switching generates lethal chromosome breaks in *rad52* strains of *Saccharomyces cerevisiae*. *Mol. Cell. Biol.* **1:** 522.

Weiffenbach, B., D.T. Rogers, J.E. Haber, M. Zoller, D.W. Russell, and M. Smith. 1983. Deletions and single base pair changes in the mating type locus that prevent homothallic mating type conversions. *Proc. Natl. Acad. Sci.* **80:** 3401.

Double-chain Breaks: Thinking about Them in Phage and Fungi

F.W. Stahl, D.S. Thaler, A. Kolodkin, S. Rosenberg, and E. Sampson
Institute of Molecular Biology
University of Oregon, Eugene, Oregon 97403

Double-chain breaks play important roles in the operation of all four of the generalized recombination pathways that can act on phage λ growing in *Escherichia coli*. For λ's own recombination pathway, Red, artificially introduced double-chain breaks initiate recombination in their neighborhood. The spontaneous recombination of the Red pathway can be attributed to duplex ends, as well. *E. coli* K12's Red homolog RecE and its recombinational repair pathway RecF act similarly. The principal *E. coli* pathway for conjugation and transduction, RecBC, is activated by double-chain breaks, but the resultant exchange can occur far from the break, preferentially near Chi (5'-GCTGGTGG).

In the yeast *Saccharomyces cerevisiae*, double-chain breaks provoke recombination in their neighborhood in meiosis. The breaks promote conversion of markers at and near the cut site. Recombination of flanking markers to give tetratype tetrads frequently accompanies that conversion. The induced conversions at the cut site are 4:0 rather than 3:1. One interpretation of that finding implies that double-chain breaks are not the usual initiators of meiotic exchange in *S. cerevisiae*.

Double-chain Breaks in Phage

Breaks Introduced by Restriction Enzymes. With a restriction system (carried on a plasmid), a double-chain break can be introduced at a known location in one of the two parents in a cross. Response to the cut can be monitored by comparing changes in recombination frequencies in intervals near to and far from the cut. When the Red or RecF pathways are operating, the consequence of a cut is an increase in the frequency of recombinants in marked intervals that include or neighbor the cut site. Data are conveniently and convincingly obtained in a cross of the sort:

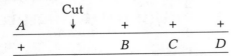

A^+D^+ recombinants in the progeny of lytic growth are selected, and the fractions of those that are recombinant in intervals AB, BC, and CD are compared. Results of such a cross, in which λ's Red pathway is responsible for recombination, are in Table 1. They demonstrate that cutting increases the fraction of recombinants in the interval containing the cut and in the interval adjacent to it at the expense of the fraction of recombinants in the interval most remote from the cut. (The yield of A^+D^+ recombinants is increased severalfold by the cutting.) Enhanced recombination in the BC interval presumably reflects recombination that follows upon degradation back from the cut site.

The view that *double*-chain breaks are responsible for the induced recombination is supported by the results of replication-blocked crosses in which the effects of cutting one parent are compared with the effects of cutting the other. The double-chain break model, in contrast with straightforward versions of single-chain break models, predicts that cutting the lower parent in the diagram above will increase the frequency of A^+D^+ recombinants in the AB interval only. Such results were obtained.

In studies like those described above for Red, *E. coli*'s RecF pathway behaves similarly. The RecBC pathway, on the other hand, behaves differently, showing a slight *decrease* in the fraction of A^+D^+ recombinants that arise in the cut interval.

cos *as a Double-chain Break Site.* Chromosomes are packaged from concatameric structures beginning at a *cos* site and proceeding rightward to a second *cos*, about 38–50 kb away. *cos* influences recombination in a manner reminiscent of restriction site stimulation (see Stahl 1986, for references). For instance, (1) only one parent

TABLE 1 Effects of *Eco*RI Cuts in Interval AB on Red-mediated Recombination (λ *red*⁺ *gam*⁺ in *rec*⁺ *E. coli*).

Interval	AB	BC	CD
Cut	0.34 + 0.03	0.24 + 0.04	0.42 + 0.05
Uncut	0.16 + 0.03	0.05 + 0.02	0.78 + 0.04

The crosses are blocked for DNA replication.

need have a cuttable *cos*, (2) Red and RecF (and RecE) recombination occur at a high rate near *cos*, but RecBC recombination does not, and (3) recombination by the former pathways is stimulated detectably in intervals 10 kb or more removed from *cos*. Unlike restriction sites, *cos* stimulates recombination primarily or exclusively to one side of itself; viz., to its left, near λ's right end. The relative inactivity of λ's left end in Red, RecF, and RecE recombination is presumably a result of the binding there of λ packaging proteins. Studies of *cos*-stimulated, Red-mediated recombination reveal details of this double-chain-break-initiated recombination reaction. A simple, well-supported model envisions 5'→3' degradation exposing a single chain overhanging at its 3' end. This 3' overhang invades an uncut duplex. The 3' end in the resulting D loop can then indulge in either of two reactions (both of which could be Pol I-mediated): (1) the 3' end primes DNA synthesis and/or (2) the 3' end provokes cutting of the displaced chain. The asymmetric behavior of *cos* (reflecting the asymmetry in DNA packaging) underlies the demonstration that the RecBC pathway is dependent on double-chain breaks, as described below.

Stimulation of RecBC Recombination by Chi. Chi (5'-GCTGGTGG) stimulates recombination in its neighborhood. It is specific to the RecBC pathway and is recognized by the wild-type *recBC* gene product (for review, see Smith and Stahl 1985). The ability of Chi to function depends on its orientation with respect to *cos*, and the ability of Chi to be recognized by *recBC* protein (as manifested by being nicked in vitro) is dependent on the direction from which the protein approaches Chi as it travels between the chains of the duplex following its entry at a duplex end. When a Chi in λ is inactive because of its incorrect orientation vis-a-vis *cos*, an *Eco*RI cut delivered at a remote place on λ activates the Chi. Other experiments imply that not only Chi-stimulated recombination but most of the remaining RecBC recombination in λ (perhaps stimulated by Chi-like sequences) is similarly dependent on a double-chain end.

Spontaneous Recombination Not Stimulated by cos. In the Red, RecE, and RecF pathways, recombination remote from *cos* is dependent on DNA replication. This dependency has been interpreted in each of two ways: (1) replication generates recombinogenic structures or (2) recombination proceeds via a break-copy pathway. Although neither possibility has been eliminated, economy of hypothesis tilts us toward (1) (Stahl et al. 1985). Economy likewise inclines us to the view that duplex ends of rolling circle tails are the postulated recombinogenic structures. Experimental support for this inclination is provided in two ways. (1) The *recBC* mutants (*recBC*‡) described by Chaudhury and Smith (1984) allow rolling circle replication. Recombination mediated by *recBC*‡ is high near a cut restriction site and near *cos* in replication-blocked crosses, whereas recombination remote from *cos* is replication dependent. (2) The *gam* protein of phage Mu binds to DNA duplex ends, and replication-dependent, Red-mediated recombination is blocked by Mu *gam* function in vivo.

Double-chain Breaks in Yeast

The *HO* nuclease makes double-chain cuts at *MAT*. These cuts presumably provoke the recombinational transfer of information from the silent loci resulting in a switch of mating type. In the absence of the silent loci, such cuts provoke mitotic conversion at *MAT* with occasional accompanying exchange of flanking markers (Klar and Strathern 1984). These results prompted an assessment, by similar methodology, of the consequence of double-chain breaks delivered at *MAT* to cells undergoing meiosis (A. Kolodkin et al., in prep.). If it is possible to stimulate recombination that is indistinguishable from spontaneous meiotic recombination, one would have support for the idea that meiotic recombination is normally initiated by double-chain breaks. In many respects the double-chain-break-induced recombination is like spontaneous meiotic recombination. However, induced conversion at the cut sites, instead of appearing as 3:1 tetrads, invariably appears as 4:0 tetrads. That result suggests that the induced recombination occurs at the two-strand stage. However, the induced crossovers accompanying the conversions are almost all tetratypes, arguing that the events are induced at the four-strand stage. A variety of resolutions of this paradox can be entertained. An economical one supposes that observed events occur in the four-strand stage but only in cells in which both cuttable chromatids are cut simultaneously by the experimentally induced *HO* nuclease. In those cells in which a cut chromatid has an uncut sister, the provoked event occurs between sisters and is, consequently, undetected. This point of view, which is harmonious with conclusions reached from radiation-induced mitotic recombination, suggests that double-chain breaks are not the normal initiators of meiotic recombination in *S. cerevisiae*, at least not in a simple sense. However, the observation by A. Klar (in prep.) of 3:1 conversions induced by double-chain breaks in *Schizosaccharomyces pombe* warns us not to close the case quickly. Hastings (1984) has raised the possibility that in *S. cerevisiae* double-chain-break repair follows simultaneous exision on the two chains of a heteroduplex that arose via an Aviemore-like reaction. By showing that meiotic *S. cerevisiae* cells deal with double-chain breaks recombinationally, we have provided comfort to all meiotic recombination models that feature such breaks.

REFERENCES

Chaudhury, A.M. and G.R. Smith. 1984. A new class of *Escherichia coli* *recBC* mutants: Implications for the role of RecBC enzyme in homologous recombination. *Proc. Natl. Acad. Sci.* **81:** 7850.

Hastings, P.J. 1984. Measurement of restoration and conversion: Its meaning for the mismatch repair hypothesis of conversion. *Cold Spring Harbor Symp. Quant. Biol.* **49:** 49.

Klar, A. and J.N. Strathern. 1984. Resolution of recombination intermediates generated during yeast mating type switching. *Nature* **30:**744.

Smith, G.R. and F.W. Stahl. 1985. Homologous recombination promoted by Chi sites and RecBC enzyme of *Escherichia coli*. *Bioessays* **2:** 244.

Stahl, F.W. 1986. Roles of double strand breaks in generalized genetic recombination. *Prog. Nucleic Acid Res. Mol. Biol.* **33:** (in press).

Stahl, F.W., I. Kobayashi, and M.M. Stahl. 1985. In phage λ, *cos* is a recombinator in the Red pathway. *J. Mol. Biol.* **181:** 199.

The Role of *RAD50* in Meiotic Intrachromosomal Recombination

J. Wagstaff, S. Gottlieb, and R. Easton Esposito
Department of Molecular Genetics and Cell Biology
The University of Chicago, Chicago, Illinois 60637

We have recently initiated a systematic study of the genetic control of recombination within single unpaired chromosomes during yeast meiosis. The experimental system that we are using in this analysis involves three essential components:

1. Haploid yeast strains that express both alleles of the *MAT* locus, either because they are disomic for chromosome III or because they contain the *mar1* mutation or because they contain a centromere plasmid carrying a *MAT* allele.
2. The *spo13-1* mutation, which eliminates reductional chromosome segregation at meiosis I and permits completion of a single meiosis II-like division.
3. Repeated genes, either naturally occurring tandem arrays such as the ribosomal DNA (rDNA) or artificially constructed repeats, for monitoring exchange.

Using this haploid meiotic system, we have found that the *SPO11* gene, which is required for meiotic exchange between homologs, is also essential for meiotic intrachromosomal recombination both between rDNA repeats and between duplicated *his4* genes on chromosome III (Wagstaff et al. 1985).

In the experiments described in this report, we have investigated the role of the *RAD50* gene in meiotic intrachromosomal recombination. Mutant strains lacking *RAD50* function are extremely sensitive to ionizing radiation (Game and Mortimer 1974). *rad50-1* diploids are defective in meiotic gene conversion and reciprocal exchange between homologs by the criteria of both interrupted meiosis experiments (Game et al. 1980) and *spo13-1* single-division meiosis (Malone and Esposito 1981).

RAD50 Is Required for Intrachromosomal Recombination between Duplicated *his4* Genes

The first genetic system used in this study contains duplicate copies of the *his4* locus on chromosome III separated by pBR313 plasmid DNA. The two *his4* loci contain different noncomplementing mutations in the *his4A* and *his4C* regions (Jackson and Fink 1981). Previous studies have shown that, during Rec$^+$ haploid meiosis, the frequency of detectable intrachromosomal recombination—unequal sister chromatid exchange (SCE), intrachromatid exchange, or gene conversion—is approximately 30%. This frequency is much higher than the 1–2% reported in diploid meiosis by Jackson and Fink (1985). The high level of exchange in the haploid meiotic system led us to conclude that the presence of the other homolog in diploid meiosis suppresses events within a single chromosome. This finding suggests that events between and within chromosomes not only use some common gene functions, such as *SPO11*, but also occur in competition with one another.

As with the *spo11-1* mutation, introduction of the *rad50-1* amber mutation into *mar1 spo13-1* strains containing this gene duplication produced approximately a 100-fold reduction in the rate of intrachromosomal recombination per meiosis (Table 1). The wild-type *RAD50* gene is thus required for both meiotic recombination between homologs and intrachromosomal recombination in a *his4* duplication. In Rec$^+$ haploids, approximately 80% of the observed events are either unequal SCEs or intrachromatid exchanges, and the remaining events consist of gene conversion or multiple exchanges; the one event detected in a *rad50-1* haploid was an intrachromatid exchange.

Table 1 Effect of *rad50-1* on Intrachromosomal Recombination in a *his4* Duplication

Genotype	Strain	Rec events/meiosis(%)
RAD50	JW201-4B	30
	JW211-15C	23
rad50-1	JW211-43B pSS31(4)	0.3

In the absence of recombination, dyads contain 2 His$^-$ spores that produce His$^+$ papillae. Recombination events during meiosis yield dyads that contain either one His$^+$ spore and one His$^-$ spore that produces His$^+$ papillae or one His$^-$ spore that produces His$^+$ papillae and one His$^-$ spore that does not.

Meiotic Intrachromosomal Recombination Within the rDNA Array Is Largely *RAD50* Independent

In contrast to the nearly complete abolition of meiotic intrachromosomal exchange between duplicated *his4* genes in *rad50-1* mutants, recombination events within the rDNA occur at frequencies that are reduced at most by two- to threefold in the absence of *RAD50* function (Table 2). We assayed exchange using strains that contained the *URA3* gene integrated in the rDNA array (Zamb and Petes 1981). Unequal SCE or intrachromatid exchange between repeats flanking the *URA3* insert yields dyads containing one Ura$^+$ spore and one Ura$^-$ spore. Although there is significant interstrain variability, the observed frequencies suggest that *RAD50* function is dispensable for a large proportion of intrachromosomal events. The source of interstrain variability may be the number of copies of *URA3* inserted into rDNA or other genetic background differences. Most events that occur in the rDNA of haploid Rec$^+$ strains are unequal reciprocal SCE (Wagstaff et al. 1985).

A Cloned *RAD50* Gene Restores *his4* Recombination in *rad50-1* Meiosis

We have isolated a 10-kb fragment of yeast DNA from a YRp17 bank (constructed by R.T. Elder) that complements the methylmethanesulfonate-sensitive phenotype of *rad50-1* strains. The frag-

Table 2 Effect of *rad50-1* on Intrachromosomal Recombination in the rDNA Array

Genotype	Strain	Rec events/meiosis(%)
Three copies of *URA3* in rDNA		
RAD50	K355-13A	1.7
	JW165-19D	6.6
	JW165-37D	1.7
	JW167-171B	1.8
rad50-1	JW168-23A	1.8
	JW168-62B	1.1
One copy of *URA3* in rDNA		
RAD50	JW226-28B	10.2
rad50-1	JW211-43B pSS31(4)	3.3

The frequency of recombinant dyads is the number of dyads with one Ura$^+$ spore and one Ura$^-$ spore divided by the number of dyads with either one or two Ura$^+$ spores.

ment integrates at the *RAD50* locus and its restriction map overlaps that reported by Kupiec and Simchen (1984). Transformation of the *rad50-1* haploid JW211-43BpSS31(4) with plasmid pSG(RAD50) yielded a Rad⁺ haploid that showed a *his4* recombination frequency of 0.25, the same as that observed in Rec⁺ haploids, and a frequency of detectable recombination events in the rDNA of 0.22 per meiosis, somewhat higher than that observed in other Rad⁺ strains.

Recombination in a *his3* Duplication, Inserted Either at the *HIS4* Locus or in the rDNA, Is *RAD50* Dependent

The results above indicate that *RAD50* plays an essential role in intrachromosomal recombination between duplicated *his4* genes as it does in meiotic exchange between homologs. However, its function is to a large extent dispensable for intrachromosomal exchange in the rDNA array.

How can we explain this dramatic difference in *RAD50*'s role in these two regions of the genome? One possibility is that specific DNA sequences within individual rDNA repeats initiate exchange in the absence of *RAD50*. Another hypothesis is that the structure and location of the rDNA, a series of tandem repeats within the nucleolus, are responsible for the unique properties of exchange characteristic of the region. To address these possibilities, we have inserted a duplication of 5′ and 3′ truncated *his3* sequences constructed by M. Fasullo and R. Davis (in prep.) at both the *HIS4* locus and in the rDNA (Fig. 1), using plasmids kindly provided by S. Roeder, and we have measured the frequency of exchange in this system in both wild-type and *rad50-1* backgrounds (Table 3). In Rad⁺ strains, the frequency of exchange yielding His⁺ products is the same at both locations, approximately 1×10^{-3} events per meiosis. For both locations, the *rad50-1* mutation causes a near-complete loss of meiotic intrachromosomal exchange. Therefore, unlike exchange between rDNA repeats, which is largely independent of *RAD50*, exchange between unique sequences inserted in the rDNA is *RAD50*-dependent.

DISCUSSION

Recombination within chromosomes can eliminate variant forms of genes from repeated families, produce new variants by intragenic exchange, and expand or contract the size of repeated families. The haploid meiosis system facilitates analysis of the gene products involved in this recombination and, by permitting examination of intrachromosomal recombination in the absence of homolog pairing,

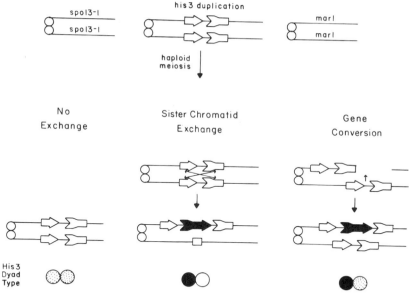

Figure 1 Recombination between repeated *his3* sequences. The 3' end of the *HIS3* gene is represented by an arrowhead; the 5' end is represented by a tail of an arrow. The two sequences overlap by 300 bp (M. Fasullo and R. Davis, in prep.). Unequal SCE between the overlap sequences or gene conversion can generate an intact *HIS3* gene, represented by a darkened arrow.

allows us to deduce the effects of homolog interactions on intrachromosomal events.

Three major conclusions emerge from the work described here:

1. Intrachromosomal exchange between repeated *his3* and *his4* sequences requires *RAD50*.
2. Intrachromosomal exchange between rDNA repeats does not show an absolute requirement for *RAD50*, although the presence of *RAD50* may stimulate this form of exchange by a factor of 2–3.
3. Intrachromosomal exchange between repeated *his3* sequences occurs with equal frequency whether the repeats are located within the rDNA array or at a non-rDNA site, and exchange is strongly *RAD50*-dependent in both locations.

These results suggest the operation of a *RAD50*-independent pathway of meiotic intrachromosomal recombination in the rDNA array, whose action may require either specific rDNA sequences or the physical arrangement of large numbers of tandemly repeated

Table 3 Recombination in a *his3* Duplication Integrated at the *HIS4* Locus or in the rDNA Array

Genotype	Number of cultures	Mean His+/cfu (× 10⁻⁴)		
		mitotic	meiotic	meiotic − mitotic
his3 integrated at *HIS4*				
rad50-1	5	1.06	1.50	0.44
rad50-1 pSG(RAD50)	5	0.50	10.47	9.97
his3 integrated in rDNA				
rad50-1	5	2.98	2.89	−0.08
rad50-1 pSG(RAD50)	5	4.13	15.86	11.73

Meiotic His+ frequencies were determined after digestion of sporulated cultures with 1% glusulase for 4 hr. pSG(RAD50) is a centromere plasmid that contains a 10-kb DNA insert that complements *rad50* mutations.

sequences. Significantly, the effects of this pathway do not extend to other DNA sequences inserted within rDNA.

What clues do we have to the function of *RAD50* in wild-type meiotic exchange? It is required for homolog recombination and for intrachromosomal recombination in non-rDNA regions. B. Byers (pers. comm.) has examined *rad50* mutant diploids and has not detected synaptonemal complexes (SCs), structures believed to mediate homolog pairing during meiosis. These data may indicate a common dependence of homolog exchange and non-rDNA intrachromosomal exchange on the presence of SCs; alternatively, *RAD-50*'s contribution to intrachromosomal exchange may be distinct from its role in SC formation. The *RAD50*-independence of rDNA exchange is intriguing in light of Byers and Goetsch's observation (1975) that SCs cannot be visualized in the nucleolus of diploid yeast cells, where most or all rDNA sequences are located.

The *spo11-1* mutation, in contrast to *rad50-1*, does not affect SC formation and sharply reduces intrachromosomal recombination in both rDNA and non-rDNA regions (Table 4). These phenotypes suggest a working model for the interaction of *RAD50* and *SPO11* in production of intrachromosomal recombinants during haploid meiosis (Fig. 2). The essential features of this model are: (1) There are at least two pathways leading to recombination between homologous DNA sequences located on a single chromosome or on homologous chromosomes. (2) One pathway functions in recombination in non-rDNA regions and the other within rDNA. (3) Analysis of the first pathway, i.e., in non-rDNA regions, indicates that exchange between homologous sequences on a single chromosome and between homologs requires at least two gene functions in common, *RAD50* and *SPO11*. This conclusion is based on the observation that the absence of either function in mutant strains reduces both types of exchange at least 100-fold, and indicates that both genes encode major *Rec* functions in meiosis. (4) The first pathway is thought to contain two branches that are both required to execute the exchange events described. One involves SC morphogenesis; the other involves the formation of appropriate enzymes to initiate and execute recombination. This conclusion is based on the observation that *rad50* mutants lack SCs while *spo11* mutants contain them. (5) The second pathway, responsible for intrachromosomal exchange between rDNA repeats, shares at least one gene function, *SPO11*, in common with recombination in non-rDNA regions, and is largely independent of the presence or absence of *RAD50*. That the *RAD50* and *SPO11* genes are unlikely to be on a single sequential dependent pathway is inferred from the fact that *SPO11* can

Table 4 Summary of Mutant Phenotypes

Genotype	Synaptonemal complexes	Meiotic recombination					
		between homologs	intrachromosomal				
			his4 duplication	rDNA array	his3 at his4	his3 in rDNA	
Wild type	+	+	+	+	+	+	
rad50-1	−	−	−	±[a]	−	−	
spo11-1	+	−	−	−	nt[b]	nt[b]	

[a]Exchange in the rDNA in rad50-1 strains is reduced up to threefold.
[b]nt, Not tested.

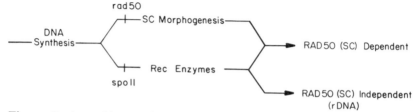

Figure 2 A working model for the interaction of *SPO11* and *RAD50* in the formation of intrachromosomal recombinants during meiosis.

function independently of the presence of *RAD50* in rDNA. In the representation in Figure 2, *SPO11* is not considered to be required for *RAD50* function because if it were, *spo11-1* mutants should lack SCs, which they do not. Tests of this model, including examination of *RAD50* and *rad50-1* haploids for SC formation and analysis of recombination involving rDNA repeats removed from the tandem array, are in progress.

REFERENCES

Byers, B. and L. Goetsch. 1975. Electron microscopic observations on the meiotic karyotype of diploid and tetraploid *Saccharomyces cerevisiae*. *Proc. Natl. Acad. Sci.* **72**: 5056.

Game, J.C. and R.K. Mortimer. 1974. A genetic study of X-ray sensitive mutants in yeast. *Mutat. Res.* **24**: 281.

Game, J.C., T.J. Zamb, R.J. Braun, M. Resnick, and R.M. Roth. 1980. The role of radiation (*rad*) genes in meiotic recombination in yeast. *Genetics* **94**: 51.

Jackson, J.A. and G.R. Fink. 1981. Gene conversion between duplicated genetic elements in yeast. *Nature* **292**: 306.

———. 1985. Meiotic recombination between duplicated genetic elements in *Saccharomyces cerevisiae*. *Genetics* **109**: 303.

Kupiec, M. and G. Simchen. 1984. Cloning and mapping of the *RAD50* gene of *Saccharomyces cerevisiae*. *Mol. Gen. Genet.* **193**: 525.

Malone, R.E. and R.E. Esposito. 1981. Recombinationless meiosis in *Saccharomyces cerevisiae*. *Mol. Cell. Biol.* **1**: 891.

Wagstaff, J.E., S. Klapholz, C.S. Waddell, L. Jensen, and R.E. Esposito. 1985. Meiotic exchange within and between chromosomes requires a common rec function in *Saccharomyces cerevisiae*. *Mol. Cell. Biol.* **5**: 3532.

Zamb, T.J. and T.D. Petes. 1981. Unequal sister-strand recombination within yeast ribosomal DNA does not require the *RAD52* gene product. *Curr. Genet.* **3**: 125.

Screening for Recombination-defective Mutants with a Positive Selection System for Plasmid Excision

R.H. Schiestl* and P.J. Hastings
Department of Genetics
University of Alberta, Edmonton, Alberta, Canada T6G 2E9

We have developed a positive selection system for intrastrand crossing-over. This was achieved by integration of a plasmid containing an internal fragment of the *his3* gene into the *HIS3* locus. It resulted in two copies of the *his3* gene, each with one terminal deletion. Exact crossing-over between the two deletion alleles restores the *HIS3* function. Conversion and mutation do not cause histidine prototrophy.

Because this selection system is present in a haploid, it is well suited to screening for mutations altering the frequency of recombination. Selection for both reciprocal products of the crossing-over event was made possible by placing an origin of replication onto the plasmid. A diploid strain was constructed that was homozygous for the integrated plasmid and included other markers for plasmid maintenance, detection of mutation, homologous conversion, and crossing-over. This strain might give clues about the mechanisms involved in plasmid excision.

A Positive Selection System for Plasmid Excision
Recombinant DNA techniques have recently been applied to create specific structures in the genome that allow the study of recombination (c.f. Orr-Weaver et al. 1981; Sugawara and Szostak 1983). We have developed a positive selection system for plasmid excision. The plasmid pRS6 (Fig. 1) used for the construction of the system contains an internal *his3* fragment with terminal deletions at both ends of the gene. In addition the *LEU2* gene is present for selection

---
*Permanent address: Department of Tumorbiology, University of Vienna, Borschkegasse 8a, 1090 Vienna, Austria.

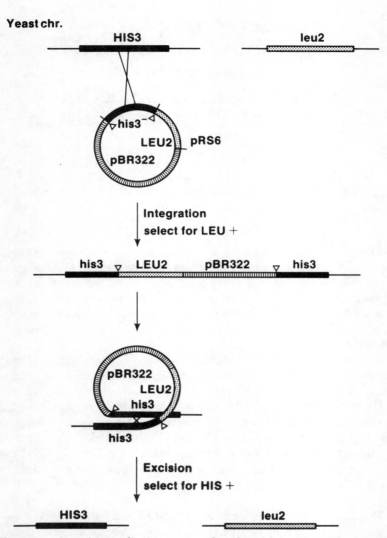

Figure 1 A positive selection system for plasmid excision. The integration of pRS6 into the *HIS3* locus and the excision of the plasmid is shown. Integration causes a change in the phenotype from *HIS⁺ leu⁻* to *his⁻ LEU⁺*. Excision reverses the phenotype again. All *HIS⁺* colonies isolated showed also a *leu⁻* phenotype.

of the plasmid in yeast and pBR322 for growth in *Escherichia coli*. The plasmid was cut with *Hin*dIII, which produced a gap in the *his3* fragment, and was then integrated into the *HIS3* locus of a histidine prototroph recipient strain (Orr-Weaver et al. 1983). The genotype of the recipient strain was changed from leu^- HIS^+ to LEU^+ his^- (Fig. 1). Southern blotting was used to screen for single integration events. Positive selection for HIS^+ yielded excision of the plasmid by crossing-over involving the two his^- genes (Fig. 1). All HIS^+ colonies tested showed a leu^- phenotype. Mutation as well as conversion is excluded from being the cause of HIS^+ prototrophy. Preselection on $-LEU$ medium abolishes background problems and the reproducibility is excellent. Plasmid excision occurs spontaneously and is found about once per 10^3 cells when selection for LEU^+ is released. Excision is inducible by UV- and x-irradiation.

Screening for Mutants Affecting RAD52-independent Recombination

Some recombination events are $RAD52$ dependent, others are independent. To understand the mechanism of $RAD52$-independent recombination, mutants that block this pathway would be of great value. The positive selection system for plasmid excision is well suited to screening for this type of mutant because recombination takes place in a haploid at a reasonably high frequency. The integrated plasmid was crossed into a $rad52^-$ background. Plasmid excision in a $rad52^-$ strain takes place spontaneously at a frequency of about 2×10^{-5}. The frequency is UV inducible at low survival levels. Cells were irradiated with UV and allowed to form colonies. Colonies that showed a much reduced frequency of spontaneous plasmid excision were isolated. Characterization will proceed by crossing strains having reduced excision frequency with a leu^- HIS^+ $RAD52$ strain.

It was shown that $RAD52$-independent conversion frequently results in chromosomal loss (Haber and Hearn 1985). To study this effect in our system, a screen for the reciprocal products of crossing-over was developed. The 2-micron ORI was placed onto the plasmid. Selection for histidine and leucine prototrophy simultaneously yields excision as well as maintenance of the plasmid. Plasmid DNA was isolated from HIS^+ LEU^+ yeast colonies and was used to transform $E.$ $coli$. Most of the yeast isolates yielded plasmid able to transform $E.$ $coli$, indicating that free plasmid was present in the yeast. Unfortunately, the frequency of HIS^+ LEU^+ cells, even in the wild type, was 200-fold lower than the plasmid excision frequency.

Plasmid Excision in Relation to Other Well-studied Genetic Events

A diploid strain homozygous for the integrated plasmid has been constructed. In addition this strain is heteroallelic for $arg4$ and $trp5$

(to measure intragenic recombination) and for *ade2* (to measure crossing-over) and homozygous for *ilv1-92* (to measure mutation) and for *ura3-52*. Evaluation of the induction kinetics with different DNA-damaging agents as well as double selection studies might permit comparison of the different genetic events.

ACKNOWLEDGMENTS

Jack Szostak and Helmut Ruis are thanked for providing very useful plasmids and yeast strains. This work was sponsored in part by a grant of the National Sciences and Engineering Council of Canada to P.J.H. and by a postdoctoral fellowship of the Alberta Heritage Foundation for Medical Research, No. 3498 to R.H.S.

REFERENCES
Haber, J.E. and M. Hearn. 1985. RAD-52 independent mitotic gene conversion in *Saccharomyces cerevisiae* frequently results in chromosomal loss. *Genetics* **111**: 7.
Orr-Weaver, T.L., J.W. Szostak, and R.J. Rothstein. 1981. Yeast transformation: A model for the study of recombination. *Proc. Natl. Acad. Sci.* **78**: 6354.
―――. 1983. Genetic application of yeast transformation with linear and gapped plasmids. *Methods Enzymol.* **101**: 228.
Sugawara, N. and J.W Szostak. 1983. Construction of specific chromosomal rearrangements in yeast. *Methods Enzymol.* **101**: 269.

Genetic and Molecular Analyses of Recombination Using Mutants Altered in DNA Repair and Sister Chromatid Recombination

M.A. Resnick,*† A.M. Chaudhury,* and J.L. Nitiss*†

*Yeast Genetics/Molecular Biology Group
Cellular and Genetic Toxicology Branch
National Institute of Environmental Health Sciences
Research Triangle Park, North Carolina 27709

†Biology Department, Illinois Institute of Technology
Chicago, Illinois 60616

Chromosomal recombination in yeast can be distinguished according to mitotic or meiotic stage of the life cycle and whether it is inter- or intrachromosomal (including sister chromatid events). Our group is interested in the genetic, molecular, and enzymatic aspects of recombination. We are determining which functions are unique and which are common to the different types of recombination (Game et al. 1980; Chow and Resnick 1983; Resnick et al. 1984, 1986), and we are assessing the role that recombination plays in other DNA metabolic processes such as repair (Game 1983; Resnick 1986). The RAD52 epistasis group and newly isolated mutants for spontaneous sister chromatid recombination (SCR) are two categories of mutants that exhibit altered recombination and are being examined extensively.

The rad52, -50, and -57 mutants exhibit a common mitotic phenotype; they are defective in the repair of radiation-induced double-strand breaks (DSBs) in chromosomes (summarized in Game 1983; J. Nitiss and M.A. Resnick, unpubl.). However, these mutants have very different phenotypes during meiosis (summarized in Table 1). Using modified return-to-mitotic medium (RMM) methods (Sherman and Roman 1963; Resnick et al. 1986), we have found that rad52 mutants exhibit as much as 10% of the wild-type levels of recombination, indicating that while the mutants are defective in recombination, meiosis can induce events that are processed in mitotic

Table 1 *rad* Mutants and Effects on Meiosis

	Rad+	rad50	rad52	rad57 34°	rad57 23°
Mitotic DSB repair	+	−	−	±	∓
Meiosis					
Viable spores	+	−	−	∓	−
Recombination in					
surviving spores	+	−	−	+	−
RMM responses					
Recombination	1	0	0.1	0.2–0.5	~1
Stable in YEPD	yes	−	no	yes	no
Time of death	−	Mei I[a]	rep/rec[b]	Mei I	rep/rec
Single strand breaks					
(per cell)	<10	<10	200–400	200–400	200–400
Time of SSBs	−	−	rep/rec	rep/rec	rec[c]
Double strand breaks					
(per cell)	<10	<10	<10	20–50	20–50

[a] Mei I, dies at about the time of meiosis I.
[b] rep/rec, occurs at a time when DNA replication and recombination are not clearly separable.
[c] rec, occurs at the time of commitment to recombination.

cells, even in the absence of *RAD52* gene product. The *rad57* mutants, which are cold-sensitive for mitotic DNA repair, exhibit meiotic lethality at both 23°C and 34°C. They have nearly RAD+ frequencies of gene conversion at both temperatures in RMM experiments; the recombinants are stable at 34°C and unstable at 23°C. The kinetics of cell death differ in that at 23°C commitment is coincident with the onset of RMM recombination, similar to results with *rad52*, while at 34°C lethality coincides with meiosis I. The *rad50* mutants do not exhibit any recombination in RMM experiments and they die at a time corresponding to meiosis I division.

These results with recombination correlate with changes in chromosomal DNA during meiosis. In *rad52* mutants, single-strand breaks (SSBs) appear when RMM recombination is detected and the final number of SSBs approximately equals the number of recombinational events in RAD+ strains (Resnick et al. 1981, 1984, 1986). No strand breaks are seen in *rad50* mutants. SSBs are also seen in the *rad57* strains at both temperatures. On the basis of their frequency and time of appearance in relation to meiotic recombination in the *rad52* and *rad57* mutants and their absence in *rad50*, we propose that these correspond to intermediate structures in meiotic recombination. These results for SSBs correlate well with recent observations of recombinant molecules in meiosis (Borts et al. 1986).

The inability to complete meiosis successfully in *rad57* at the "permissive" temperature for mitotic DNA repair could be due to a limited ability for dealing with the much larger number of recombination-initiating events that occur in meiosis as compared with following radiation treatment. The loss of survival in *rad57* at the time of meiosis I when cells are kept at the "permissive temperature" may indicate that reciprocal recombination, which is required for the proper segregation of chromosomes, is not completed.

It is interesting that although the above mutants were originally isolated on the basis of X-radiation sensitivity, they subsequently have been shown to have marked differences in meiotic behavior. Although they lacked DSB repair in mitotic cells, there had been no clear physical evidence of DSBs during meiosis. Recently we found a small number of DSBs appearing in the *rad57* mutants. Their frequency, time of appearance, and distribution are different from the SSBs; possibly they arise from a processing of the SSBs.

Recombination not only plays an important biological role in providing a mechanism for proper chromosomal segregation in meiosis, but it is also essential in mitotic cells for DNA repair. Sister chromatid recombination (SCR) has long been implicated in repair, particularly for DSBs induced in haploids (Brunborg et al. 1980). We have analyzed several *rad* mutants that previously had been characterized for effects on recombination between homologous chromosomes. To understand SCR further, we have also isolated mutants in this process. These mutants are useful for evaluating not only the relationship between sister chromatid and homologous chromosome recombination, but they also might provide a unique genetic approach for analyzing DNA replication. We have utilized the construction of M. Fasullo and R. Davis (in prep.) for detecting SCR; it consists of a rearranged split gene (*HIS3*) with a region of homology (300 bp; indicated by 0000):

5'···0000----˃····˃----0000···3'
where the complete gene is
5'····˃----0000----˃····3'

Alignment of homologous regions in sister chromatids followed by recombination will generate a functional gene.

The rates of SCR His⁺ recombinants generated in haploid and diploid cells are 7.4×10^{-6} and 3.9×10^{-6}, respectively. There is no increase in SCR during meiosis, although recombination between the construction on homologous chromosomes does increase. Ultraviolet light causes a dose-related increase at nonlethal doses. The results for ionizing radiation, for which there is no dose-related increase, could be due to physical constraints on SCR during G-2 with the construction we have used, which is near the centromere.

We have examined various *rad* mutations for their effects on SCR (summarized in Table 2). The spontaneous SCR levels of *rad52* and *rad54* (at the restrictive temperature) are low compared with Rad⁺, implicating the corresponding gene products in SCR. The results of Zamb and Petes (1981) in which SCR in the rDNA was normal for *rad52* may be due to different mechanisms for recombination in rDNA. The *rad50* mutants exhibit a moderate reduction in the spontaneous level and *rad18* is comparable to the wild type; *rem1*, which is an allele of *rad3* (R. Malone, pers. comm.), has enhanced SCR.

Mutants for SCR were isolated after ethylmethanesulfonate (EMS) treatment followed by the identification of colonies that showed either elevated (*esr*) or decreased (*dsr*) levels of SCR. Among the *esr* mutants, 10- to 15-fold increases in rates over wild type are observed; for the *dsr* mutants the rates were reduced approximately 10-fold. Three complementation groups have been identified for *esr* mutations and two for *dsr*. The *esr1* mutant is somewhat UV and X-ray sensitive and enhances the appearance of can^R mutants; it also shows heightened mitotic gene conversion. The *esr2* and *esr3* and the *dsr* mutants do not exhibit these sensitivities or a mutator phenotype.

To investigate the generality of the SCR results and the role of the ESR genes, we are determining the effects of these mutations on recombination in CEN plasmids. An SCR event would generate a disomic structure which would be unstable. Unlike the situation with wild-type strains, where CEN plasmid transformants are stable, nearly all the transformants of *esr1* exhibit low levels of stabil-

Table 2 Sister Chromatid Recombination in DNA Repair and in SCR Mutants

Mutants	Relative rate[a]
RAD⁺, *rad1*, *rad18*	+ +
rad50	+
rad54	
(24°C)	+ +
(34°C)	−
esr1, *esr2*, *esr3*, *rem1-1*	+ + +
dsr1, *dsr2*, *rad52*	−

[a] + + +, $>50 \times 10^{-6}$ events per cell generation; + +, $5\text{--}10 \times 10^{-6}$; +, $1\text{--}5 \times 10^{-6}$; −, $<1 \times 10^{-6}$.

ity (comparable to YRP plasmid transformants). These transformants of the *esr1* strains are being analyzed further.

We conclude from the results with the *rad*, *esr*, and *dsr* mutants that gene products required in mitotically identified DNA repair can be essential in several types of recombination; processes involved in both DNA repair and homologous chromosome recombination can also function in SCR. Using these mutants we have been able to analyze at the genetic and molecular level various aspects of mitotic and meiotic recombination between homologous chromosomes and sister chromatids. The isolation of the *esr* and *dsr* mutants will facilitate investigations into the role of SCR in DNA repair and in gene duplication and amplification. The SCR mutants that are not sensitive to DNA-damaging agents suggest that some genes associated with the SCR may function in processes other than DNA repair.

ACKNOWLEDGMENT

We wish to thank M. Fasullo for providing the plasmid containing the *his3* split gene construction and the strain with the *HIS3* deletion.

REFERENCES

Borts, R.H., M. Lichten, and J.E. Haber. 1986. Analysis of meiosis-defective mutations in yeast by physical monitoring of recombination. *Genetics* (in press).

Brunborg, G., M.A. Resnick, and D.H. Williamson. 1980. Cell cycle specific repair of DNA double-stranded breaks in *Saccharomyces cerevisiae*. *Radiat. Res.* **82**: 547.

Chow, T.Y.-K. and M.A. Resnick. 1983. The identification of a deoxyribonuclease controlled by the *RAD52* gene of *Saccharomyces cerevisiae*. In *Cellular responses to DNA damage* (ed. E.C. Friedberg and B.A. Bridges), p. 447. A.R. Liss, New York.

Game, J.C. 1983. Radiation sensitive mutants and repair in yeast. In *Yeast genetics* (ed. J.F.T. Spencer et al.), p. 109. Springer-Verlag, New York.

Game, J.C., T.J. Zamb, R.J. Braun, M. Resnick, and R.M. Roth. 1980. The role of radiation (*rad*) genes in meiotic recombination in yeast. *Genetics* **94**: 51.

Resnick, M.A. 1986. Genetic control of meiotic recombination and biochemical molecular events. In *Meiosis* (ed. P. Moens). Academic Press, New York. (In press.)

Resnick, M.A., T. Chow, J.L. Nitiss, and J.C. Game. 1984. Changes in chromosomal DNA of yeast during meiosis in repair mutants and the possible role of a deoxyribonuclease. *Cold Spring Harbor Symp. Quant. Biol.* **49**: 639.

Resnick, M.A., J. Nitiss, C. Edwards, and R. Malone. 1986. Meiosis can induce recombination in *rad52* mutants of *Saccharomyces cerevisiae*. *Genetics* (in press).

Resnick, M.A., J.N. Kasimos, J.C. Game, R.J. Braun, and R.M. Roth. 1981. Changes in DNA during meiosis in a repair-deficient (*rad52*) of yeast. *Science* **212:** 543.

Sherman, F. and H. Roman. 1963. Evidence for two types of allelic recombination in yeast. *Genetics* **48:** 255.

Zamb, T.J. and T.D. Petes. 1981. Unequal sister-strand recombination within yeast ribosomal DNA does not require the *RAD52* gene product. *Curr. Genet.* **3:** 125.

A Site for the Initiation of Gene Conversion in Meiosis

R.E. Malone,* S. Cramton, and R. Gehrhardt
Department of Microbiology, Stritch School of Medicine
Loyola University of Chicago, Maywood, Illinois 60153

The properties of gene conversion in fungi strongly suggest that there are specific sites or regions where genetic exchange starts (Holliday 1964; Murray 1968), if, as seems likely, gene conversion events represent a "signal" for the initiation of reciprocal exchange (Fogel et al. 1979). The most obvious property is the polarity of gene conversion—the decrease in the amount of gene conversion from one side of the gene to the other. The patterns of coconversion of multiple alleles within a locus are also consistent with the hypothesis (Fogel et al. 1981). There are a number of mutations known that increase the frequency of gene conversion in their vicinity and have been interpreted as possible initiation sites. For example, the M26 mutation in the ade6 gene of Schizosaccharomyces pombe stimulates conversion an order of magnitude, and also affects coconversion of other ade6 alleles in a manner consistent with recombination starting at its position. One of the most intriguing mutations is the cog mutation in Neurospora crassa (Catcheside and Angel 1974). It can stimulate recombination 10-fold or more, even when heterozygous. Furthermore, it can stimulate recombination across a region of heterology, if it is present on the chromosome without the heterology.

Many of the interpretations of the genetic data obtained from studying gene conversion in yeast are dependent upon a model for recombination, such as a strand invasion model (Meselson and Radding 1975) or a double-strand-break model (Szostak et al. 1983). One phenomenon that seemed to us to be relatively independent of any model was the dependence of conversion upon distance, such as that displayed by polarity. An examination of gene conversion at various loci in Saccharomyces cerevisiae indicates a wide range of frequencies ranging from 0.6% to 18% (Fogel et al. 1979). We felt

*Present address: Department of Biology, University of Iowa, Iowa City, Iowa 52242.

that this probably reflected the distance that the gene was located from an initiation site. The available data (such as the length of coconversion tracts) are consistent with the hypothesis that meiotic conversion tracts extend over several hundred to no more than several thousand bases. We therefore felt that if we cloned a locus that displayed a high frequency of conversion, as well as several thousand base pairs on either side, it should be possible to isolate an initiation site and study it.

RESULTS

The *his2* locus was reported by Fogel et al. (1979) to have a 18% frequency of gene conversion (including all types of abberrant segregation). We therefore cloned the *HIS2* gene and some 4 kb on either side. Using the cloned fragment, we could target integration at the *HIS2* locus and could show linkage between inserted genes and *HIS2*. Our approach to finding the putative initiation site was based upon two concepts. First, since conversion is dependent upon distance, an insertion of DNA between the initiation site and the gene monitored should lower the frequency; an insertion on the other side of the initiation site or on the other side of the gene should have little or no effect. Second, analysis of the insertions would be done by placing the insertions into the chromosome and analyzing unselected tetrads. In this way we would avoid any bias due to mismatch repair (which might be encountered when selecting His$^+$ prototrophs from two *his2* heteroalleles), or due to properties of extrachromosomal plasmids (which might have different rules for recombination than chromosomes).

The data in Table 1 indicate that recombination at the *HIS2* locus displays normal properties expected for gene conversion, such as parity, and a 56% association with crossing-over. Like Fogel et al. (1979), we observed little postmeiotic segregation, although only 150 of the tetrads were tested by directly replicating the spore clones. The frequency of conversion in our laboratory strains ranges from 12% to 14%, which is significantly less than the 18% reported by Fogel et al. (1979), even though we are using the same allele (*his2-1*). We attribute this to strain differences, but have not pursued it further because the frequency is consistent within our strains.

A map of the *HIS2* region and the location of some of the DNA fragments inserted in the region is shown in Figure 1. We have used two types of insertions in the hope that we could avoid being misled by effects caused by the insert other than its size. (More recently, we have made a third series of insertions based on Fred Heffron's "shuttle mutagenesis" vectors. These are currently being analyzed.)

Table 1 Properties of Gene Conversion Occurring at the HIS2 Locus

Diploid genotype	CDC14	HIS2	+
	+	+	CLY3

Number tetrads	Conversion patterns				Total	%
	3:1	1:3	4:0	0:4		
407	26	21	2	1	50	12.4

	Map distances		
	CLY3-HIS2	HIS2-CDC14	CLY3-CDC14
Experimental	6.4	4.6	13.0
Published[a]	5.2	(0.4-6.1)	ND

Association of conversion with exchange	
Uncorrected for incidental exchange:	64%
Corrected for incidental exchange:	56%

[a]Mortimer and Schild (1980).

Analysis of crosses containing insert URA3#1 (to the left of the HIS2 gene) indicate that the insert does not diminish the amount of gene conversion when present in either the hetero- or homozygous state (Table 2). In fact, conversion is stimulated somewhat. This has been observed for insertions of URA3 at the MAT locus (J. Haber, pers. comm.). It is interesting to note, however, that 5/12 HIS2 conversions coconverted the URA3 insert when it was present in a heterozygous state. No URA3 conversions alone were observed. This is

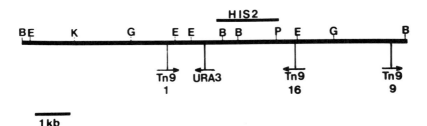

Figure 1 Map of HIS2 region and location of insertions studied for effect on recombination. The position of the HIS2 gene has been determined from subclones, ExoIII deletion mapping, and insertion mutagenesis. (B) BamHI; (E) EcoRI; (G) BglII; (K) KpnI; (P) PstI.

Table 2 Gene Conversion at *HIS2* in the Presence of Various Insertions

Genotype	Frequency conversion	Number	Type of tetrad				comments
			3:1	1:3	4:0	0:4	
his2-1	12.4%	50/407	26	21	2	1	associated
HIS2							c.o. = 56%
his2-1	14.5%	12/83	6[a]	6[b]	0	0	
HIS2::URA3#1							
his2-1::URA3#1	19.8%	47/232	24	21	1	1	
HIS2::URA3#1							
his2-1	13.3%	6/45	3	3	0	0	
HIS2::Tn9#1							
his2-1	7.4%	26/350	20	4	1	1	ratio $\frac{3:1}{1:3} = 5$
HIS2::Tn9#16							
his2-1::Tn9#16	4.2%	6/143	3	3	0	0	parity restored
his2-1::Tn9#16							

[a] 2/6 coconverted for *URA3* 3:1.
[b] 3/6 coconverted for *URA3* 1:3.

consistent with an initiating event to the right of *HIS2*. A limited number of tetrads analyzed for Tn9#1 (also located to the left of *HIS2*) give similar results. A heterozygous Tn9#1 insert does not reduce conversion at *HIS2*. In contrast, Tn9#16, inserted about 0.4 kb to the right of the *HIS2* gene, reduces the conversion to 7.4% when heterozygous (Table 2). This would be consistent with the initiation site lying to right of Tn9#16, if the recombination event was reduced by a region of heterology. Note that parity does not exist; 3:1 tetrads exceed 1:3 tetrads by a factor of 5 (see Discussion). Finally, the homozygous Tn9#16 caused a reduction of conversion to a frequency of 4.2% (Table 2).

If the initiation site were located to the right of the Tn9#16 insert, many of the conversion events at *HIS2* should have coconverted the Tn9#16 insert when it was heterozygous. A Southern analysis of recombinant tetrads indicates that this is not the case (Fig. 2). Eight of 12 tetrads having a *HIS2* conversion segregate 2:2 for Tn9#16. The other four tetrads do not show coconversion, but, rather surprisingly, show the existence of a novel band (Fig. 2). The size of the new band is uniformly 6 kb in all tetrads; it represents the loss of the internal Tn9 portion and one IS1 element, as shown by probing with an internal Tn9 fragment (Fig. 2). We presume that this occurs by an intrachromosomal crossover event between the direct IS1 repeats. Such events can occur even when the Tn9 is homozygous (Fig. 2).

DISCUSSION

It is clear that the Tn9#16 inserts we have made in the chromosome have had an effect on gene conversion at *HIS2*. In terms of the effect on conversion frequency, the results are consistent with the existence of an initiation site to the right of the insertion site of Tn9#16. The effects on parity when the Tn9 is heterozygous are also consistent—at least with the double-strand-break model. However, when one looks closer at the DNA of the recombinant tetrads, the data are less consistent with such a location. The major difficulty is the failure to observe any coconversion of Tn9 in the tetrads examined on Southerns. Neither the strand invasion nor the double-strand-break models appear to explain readily all the observations. Perhaps the best model is one that presumes an enzyme starts at a site to the right of Tn9#16, and moves to the left. With decreasing probability over distance, it initiates the recombination event. When initiation starts in the heterozygous Tn9, it results in either an nonreproductive event (with respect to the *HIS2* locus) or an intrachromosomal event leading to the "pop-out." However, the double-

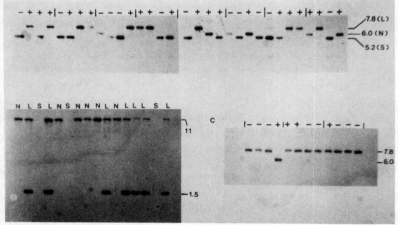

Figure 2 Analysis of DNA from recombinant and parental tetrads. (A) Analysis of DNA from tetrads displaying a gene conversion at HIS2 in the presence of a heterozygous Tn9#16. The HIS2 parent contained the Tn9, generating a Large BglII fragment of 7.8 kb. The his2 parent contained the normal (Small) band of 5.2 kb. Note the presence of a Novel band at the size of 6.0 kb. (B) DNA (digested by PvuII) from cells containing either the S, N, or L BglII band probed with a PstI Tn9 internal fragment of 1.9 kb. The complete Tn9 (in the L BglII band) generates three fragments: 11 kb, 1.5 kb, and 0.4 kb (not shown). The small (S) BglII band has no Tn9. The novel band is missing both internal 1.5- and 0.4-kb bands. From this and other digests, it is possible to infer that only one IS1 is left in the novel band. (C) DNA from a cross where the Tn9#16 is homozygous. The novel band appears again, suggesting that it is not due to just the heterozygosity of the Tn9. The + or − symbols above the lanes represent the His phenotype of the spore clone from which the DNA was made.

strand-break model can also partially explain the data obtained from the heterozygous Tn9#16, if the initiation site were located to the left of the insert, between the Tn9 and the HIS2 gene. It still does not explain why the conversion frequency falls when the insert is homozygous, if the site were to the left. It is possible that the Tn9 element itself is affecting recombination. We plan to test this by looking at other Tn9 inserts in the region, as well as inserts of yeast DNA that do not have repeats. In addition, we have constructed subclones of this region and inserted them next to the LYS2 locus, but these have not yet been inserted into the chromosome.

We feel that we have defined a region of interest for the initiation of gene conversion, and that our current data do not easily fit with either popular model for recombination in yeast. Further experiments should help to define exactly what sequences are required for conversion at *HIS2*. It is entirely possible that the initiation site will prove to be complex, such as the relative AT/GC ratio over a region of several thousand base pairs, or "higher-order" chromosome structure. Studies of a putative initiation site near *PYK1* have indicated that that site is complex also (D. Olson, pers. comm.).

ACKNOWLEDGMENTS

This work was supported by National Institutes of Health grant RO1-GM29172 to R.E.M. The authors appreciate the helpful discussions about recombination hot-spots with David Olson.

REFERENCES

Catcheside, D.G. and T. Angel. 1974. A histidine-3 mutant, in *Neurospora crassa*, due to an interchange. *Aust. J. Biol. Sci.* **27**: 219.

Fogel, S., R. Mortimer, and K. Lusnak. 1981. Mechanisms of meiotic gene conversion, or "wanderings on a foreign strand." In *The molecular biology of the yeast* Saccharomyces: *Life cycle and inheritance* (ed. J. Strathern et al.), p. 289. Cold Spring Harbor Laboratory, Cold Spring Harbor, New York.

Fogel, S., R. Mortimer, K. Lusnak, and F. Tavares. 1979. Meiotic gene conversion: A signal of the basic recombination event in yeast. *Cold Spring Harbor Symp. Quant. Biol.* **43**: 1325.

Holliday, R. 1964. A mechanism for gene conversion in fungi. *Genet. Res.* **5**: 282.

Meselson, M. and C. Radding. 1975. A general model for genetic recombination. *Proc. Natl. Acad. Sci.* **72**: 358.

Mortimer, R.K. and D. Schild. 1980. Genetic map of *Saccharomyces cerevisiae*. *Microbiol. Rev.* **44**: 519.

Murray, N. 1968. Polarized intragenic recombination in chromosome rearrangements of *Neurospora*. *Genetics* **58**: 181.

Szostak, J., T. Orr-Weaver, R. Rothstein, and F. Stahl. 1983. The double-strand break repair model for recombination. *Cell* **33**: 25.

REC Genes Governing Mitotic Recombination, Chromosomal Stability, and Sporulation: Cell Type and Life Cycle Stage-specific Expression of *rec* Mutants

M.S. Esposito, K. Bjornstad, L.L. Holbrook, and D.T. Maleas

Cell and Molecular Biology Group, Biology and Medicine Division
Lawrence Berkeley Laboratory
University of California, Berkeley, California 94720

We have isolated a broad spectrum of UV light-induced hyporecombination and hyperrecombination mutants affecting spontaneous mitotic gene conversion and intergenic recombination on chromosome VII in an $n+1$ *MATα* strain (LBL1, Table 1) disomic for this chromosome (Esposito et al. 1982, 1984; Esposito 1984). Our goal is to identify *REC* genes encoding the partial reactions comprising mitotic recombination and their effects upon meiosis. Five phenotypic classes of hyporecombination and hyperrecombination *rec* mutants were recovered (Esposito et al. 1984) and several were subsequently demonstrated to have analogous effects on in vitro plasmidic recombination in cell-free extracts (Symington et al. 1984). DNA cellulose chromatography has been used as a method to fractionate yeast DNA-processing proteins (e.g., SSB-1, SS-DNA-dependent ATP-

Table 1 Principal Strains Employed

Strain		Genotype
LBL1	*MATα*	*ade5 met13-c cyh2^r trp5 LEU1 ade6 cly8*
		ADE5 met13-d CYH2^s TRP5 leu1 ADE6 CLY8
		his7-1 tyr1-2 lys2-2 adede2-1 ura3-1 CAN1^s
NLBL1	*MATα*	*ade5 met13-c cyh2^r trp5 LEU1 ade6 cly8*
		his-7-1 tyr1-2 lys2-2 ade2-1 ura3-1 CAN1^s
NLBL3	*MATa*	*ADE5 met13-d CYH2^s TRP5 leu1 ADE6 CLY8*
		his7-2 tyr1-1 lys2-1 ade2-1 ura3-1 can1^r

ase, and resolvase) for which mutants may be defective (Hosoda et al. 1985). In this report we summarize the properties of *rec* mutants representative of each of the five phenotypic classes.

Properties of *MATα rec n + 1* Disomics

Representatives of each phenotypic class of *rec* mutants are shown in Table 2. The mutant *rec490* exhibits a temperature-sensitive hyporecombination (Rec⁻) phenotype and is defective in both gene conversion of CYH^s to cyh^r and intergenic recombination in the interval *TRP5–CYH2*. The *rec490* mutation is recessive and complements *rec754*, a *rec* mutation exhibiting the same Rec⁻ phenotype. Conversion⁻ Intergenic Recombination⁻ *rec* mutants and certain *rad* mutants (e.g., *rad52*) demonstrate that conversion and intergenic recombination are under coordinate genetic control.

The mutant *rec413* (Table 2) is defective in intergenic recombination in the intervals *TRP5–CYH2* and *MET13–ADE5*. *rec413* retains the capacity for single-site conversion of $CYH2^s$ to $cyh2^r$ and heteroallelic recombination of *MET13* heteroalleles. The *rec413* mutant, and other *rec* mutants having the same phenotype (*rec100*, *rec336*, and *rec467*) demonstrate that there are gene products required for intergenic recombination that are not required for gene conversion. Conversion⁺ Intergenic Recombination⁻ mutants may provide a means to identify genes that encode products required after the initiation of recombination.

The mutant *rec109* (Table 2) is defective in conversion of $CYH2^s$ to $cyh2^r$ and heteroallelic recombination of *MET13* alleles. Intergenic recombination in the intervals *TRP5–CYH2* and *MET13–ADE5* is observed. Conversion⁻ Intergenic Recombination⁺ mutants are not anticipated by some models of recombination. Conversion⁻ Intergenic Recombination⁺ mutants defective in *single-site* conversion are apparently rare, since *rec109* is the only mutant of this type we have recovered.

The mutant *rec193* (Table 2) exhibits an extreme hyperrecombinational (Rec⁺⁺) phenotype with respect to intergenic recombination in the *TRP5–CYH2* interval and slight stimulation of conversion of $CYH2^s$ to $cyh2^r$. Conversion⁺ Intergenic Recombination⁺⁺ mutants such as *rec193*, *rec146*, *rec395*, *rec409*, and *rec952* further demonstrate that conversion and intergenic recombination are under separate genetic control. The interpretation of this class of mutants is highly reliant on an understanding of the molecular origin of intergenic recombinants, which is still unclear.

The mutant *rec46* (Table 2) is representative of a rare class of Rec⁺⁺ mutants that exhibit an extreme Rec⁺⁺ phenotype with re-

Table 2 Conversion, Intragenic Recombination, Intergenic Recombination, and Chromosomal Loss in *MATα* Chromosome VII Disomics

Strain	Leu⁺Trp⁺Cyhʳ		Cyhʳ total	Met⁺		Number of experiments
	red	white		white	total	
LBL1	70.2×10^{-7}	137.9×10^{-7}	159×10^{-6}	6.8×10^{-8}	28.2×10^{-7}	30
rec490						
(36°C)	$\underline{0.0 \times 10^{-7}}$	$\underline{0.0 \times 10^{-7}}$	0.0×10^{-6}			5
(24°C)	116.0×10^{-7}	57.0×10^{-7}	548×10^{-6}			5
rec413	189.5×10^{-7}	$\underline{0.0 \times 10^{-7}}$	234×10^{-6}	$\underline{0.0 \times 10^{-8}}$	28.9×10^{-7}	10
rec109	$\underline{7.6 \times 10^{-7}}$	60.7×10^{-7}	891×10^{-6}	9.0×10^{-8}	$\underline{6.4 \times 10^{-7}}$	10
rec193	166.5×10^{-7}	$\underline{222.6 \times 10^{-6}}$	$\underline{123 \times 10^{-5}}$			5
rec46	$\underline{100.0 \times 10^{-6}}$	$\underline{247.0 \times 10^{-5}}$	$\underline{175 \times 10^{-4}}$			5

Median frequencies.

spect to both conversion and intergenic recombination. Conversion⁺⁺ Intergenic Recombination⁺⁺ mutants like *rec46* and *rec199* (a *cdc9* mutation) also demonstrate that conversion and intergenic recombination levels are under coordinate genetic control. Conversion⁺⁺ Intergenic Recombination⁺⁺ and Conversion⁺ Intergenic Recombination⁺⁺ mutants typically exhibit enhanced levels of chromosomal loss. This aspect of their phenotype is consistent with the view that intergenic recombination in $n+1$ disomes frequently results from incomplete (thus nonreciprocal) exchange events.

Properties of *MATa/MATα rec/rec* Diploids

Recombination and Sporulation. We have characterized *MATa/MATα rec/rec* diploid hybrids for their effects upon mitotic recombination, chromosomal stability, sporulation, and ascospore viability. One goal of these studies is to determine whether *rec* mutants isolated in a *MATα* $n+1$ disome exhibit the same phenotype in *MATa/MATα* diploid cells. We anticipated that there might be differences because *MATα* $n+1$ disomics and *MATa/MATα* diploids exhibit consistent differences in the relative rates of occurrence of conversion and intergenic recombination. Conversion of the same markers occurs in *MATa/MATα* Rec⁺ diploids at rates that are nearly equal to or at most fivefold higher than those of *MATα* Rec⁺ $n+1$ disomics, whereas intergenic recombination in *MATa/MATα* Rec⁺ diploids occurs at rates that are consistently five- to tenfold higher than those of *MATα* $n+1$ Rec⁺ strains. Thus, mitotic recombination in *MATa/MATα* diploids could be either *more or less* dependent upon *REC* gene products essential for *MATα* $n+1$ recombinational events.

We also wish to determine whether *rec* mutants that affect chromosome VII mitotic recombinational events affect mitotic recombination on other chromosomes. The eventual aim is to construct *MATa/MATα rec/rec* diploids heterozygous and heteroallelic for chromosome II, V, and VII markers. The residual genotype of these hybrids is that obtained by mating NLBL1 × NLBL3 (Table 1). Our progress to date is summarized in Table 3 and in Esposito et al. (1986).

Since our analyses involve conditionally Rec⁻ mutants, we have determined mitotic rates of recombination for the control *REC/REC* hybrid (NLBL1 × NLBL3) at both 36°C and 24°C. Intragenic recombination of *LYS2*, *TYR1*, *HIS7*, and *MET13* heteroalleles occurs at essentially the same rates at 24°C and 36°C. Single-site conversion of $CYH2^s$ to $cyh2^r$ and intergenic recombination in the intervals

Table 3 Mitotic Gene Conversion, Intragenic Recombination, and Intergenic Recombination in $MAT\alpha/MAT\alpha$ Diploids

Genotype	Temperature	Leu+Trp+Cyhr red	white	Cyhr total	Met+ white	total	Lys+	Tyr+	His+	Canr	Number of experiments
$\frac{REC}{REC}$	36°C	49.0×10^{-6}	762.0×10^{-6}	98×10^{-5}	41.0×10^{-7}	26.0×10^{-6}	20×10^{-7}	23×10^{-7}	31.0×10^{-7}	45×10^{-5}	8
$\frac{REC490}{rec490\text{-}1}$	36°C	74.0×10^{-6}	430.0×10^{-6}	73×10^{-5}	14.0×10^{-7}	16.0×10^{-6}	35×10^{-7}	41×10^{-7}	56.0×10^{-7}	23×10^{-5}	7
$\frac{rec490\text{-}1}{rec490\text{-}1}$	36°C	$\underline{13.0\times10^{-6}}$	$\underline{52.0\times10^{-6}}$	$\underline{17\times10^{-5}}$	$\underline{16.0\times10^{-7}}$	$\underline{19.0\times10^{-6}}$	14×10^{-7}	27×10^{-7}	27.0×10^{-7}	$\underline{17\times10^{-5}}$	18
$\frac{REC}{REC}$	24°C	72.0×10^{-7}	446.0×10^{-7}	88×10^{-6}	9.4×10^{-7}	10.0×10^{-6}	15×10^{-7}	11×10^{-7}	10.0×10^{-7}	34×10^{-5}	23
$\frac{REC413}{rec413\text{-}1}$	24°C	$\underline{2.0\times10^{-8}}$	$\underline{0.0\times10^{-8}}$	$\underline{3.5\times10^{-8}}$			16×10^{-7}	29×10^{-7}	9.8×10^{-7}	$\underline{89\times10^{-5}}$	11
$\frac{REC413\ sup413}{rec413\text{-}1\ SUP413}$	24°C	24.0×10^{-7}	192.0×10^{-7}	29×10^{-6}							5
$\frac{rec413\text{-}1}{rec413\text{-}1}$	24°C	$\underline{0.0\times10^{-8}}$	$\underline{0.0\times10^{-8}}$	$\underline{0.0\times10^{-8}}$					9.1×10^{-7}	23×10^{-5}	5
$\frac{REC46}{rec46\text{-}1}$	24°C	107.0×10^{-7}	819.0×10^{-7}	247×10^{-6}			13×10^{-7}		15.0×10^{-7}	30×10^{-5}	18
$\frac{rec46\text{-}1}{rec46\text{-}1}$	24°C	$\underline{71.0\times10^{-6}}$	$\underline{641.0\times10^{-6}}$	$\underline{99\times10^{-5}}$			10×10^{-6}		16.0×10^{-6}	$\underline{47\times10^{-4}}$	18

TRP5–CYH2 and ADE5–MET13 occur at rates that are five- to tenfold higher at 36°C than at 24°C.

At 36°C rec490-1/rec490-1 diploids exhibit a hyporecombinational phenotype with respect to both $CYH2^s \rightarrow cyh2^r$ conversion and intergenic recombination in the TRP5–CYH2 and MET13–ADE5 intervals. Overall heteroallelic recombination rates at MET13, LYS2, TYR1, and HIS7 are unaffected (Table 3). The rec490-1 mutation is recessive (Table 3) and confers a recessive, temperature-sensitive Spo⁻ phenotype that cosegregates with the rec490-1 mutation (Table 4). The REC490 gene is thus required in MATa/MATα cells for the class of mitotic recombinational events that are stimulated by growth at 36°C, recombinational events that occur at 36°C in MATα $n+1$ disomics and for sporulation of MATa/MATα cells.

The rec413 mutation is dominant. REC413/rec413-1 MATa/MATα diploids sporulate well but yield few (<1%) viable ascospores. A dominant suppressor of rec413-1, designated SUP413, was isolated serendipitously and employed to obtain MATa and MATα rec413-1 haploid segregants. MATa/MATα rec413-1/rec413-1 diploids are Rec⁻ with respect to both $CYH2^s \rightarrow cyh2^r$ conversion (unlike the MATα rec413-1 $n+1$ disomic strain) and Rec⁺ with respect to heteroallelic recombination at HIS7. MATa/MATα rec413-1/rec413-1 diploids are Rec⁻ with respect to intergenic recombination in the TRP5–CYH2 interval. MATa/MATα rec413-1/rec413-1 diploids sporulate well but ascospore survival is drastically reduced (Table 4).

The hyperrecombination mutant rec46-1 is recessive (Table 3) and stimulates the rates of gene conversion, intragenic recombination, and intergenic recombination approximately tenfold in all intervals tested. The enhancement of the rate of intergenic recombination by rec46-1 in MATa/MATα diploid cells is tenfold lower than that observed in the MATα rec46-1 $n+1$ disomic in which there is a 100-

Table 4 Sporulative Ability of MATa/MATα rec/rec Diploids

Strain	Temperature	Percent four- and three-spored asci	Percent two- and one-spored asci	Percent total asci
REC/REC	24°C	72.2	19.5	91.7
REC/REC	36°C	61.8	21.6	83.4
rec490-1/rec490-1	24°C	70.0	20.2	92.2
	36°C	0.3	0.3	0.6
rec413-1/rec413-1	24°C	68.4	9.2	77.6[a]
rec46-1/rec46-1	24°C	1.0	1.7	2.7

[a] Ascospore viability is less than 1×10^{-3}.

fold rate enhancement. The *REC46* gene is required for sporulation; a Spo⁻ phenotype cosegregates with the *rec46-1* mutation. Additional genetic properties of *rec46-1* are reported in Esposito et al. (1986).

Chromosomal Instability of MATa/MATα rec/rec *Diploids*. *MATa/ MATα rec/rec* diploids are being studied to determine whether they exhibit chromosomal loss. Since it appears that *MATa/MATα* $2n-1$ segregants monosomic for chromosome VII are rare or inviable, we have employed chromosome VII trisomics ($CYH2^s/cyh2^r/cyh2^r$) to monitor loss of the $CYH2^s$-bearing chromosome resulting in Cyh^r $2N$ segregants. Loss of chromosome III or chromosome V results in $2n-1$ segregants that are viable and readily detected by selective techniques.

Studies in progress demonstrate that *rec413-1* and *rec490-1* enhance chromosomal instability in *MATa/MATα* cells.

REFERENCES

Esposito, M.S. 1984. Molecular mechanisms of recombination in *Saccharomyces cerevisiae*: Testing mitotic and meiotic models by analysis of hyporec and hyper-rec mutations. *Symp. Soc. Exp. Biol.* **38**: 123.

Esposito, M.S., D.T. Maleas, K.A. Bjornstad, and C.V. Bruschi. 1982. Simultaneous detection of changes in chromosome number, gene conversion and intergenic recombination during mitosis of *Saccharomyces cerevisiae*: Spontaneous and ultraviolet light induced events. *Curr. Genet.* **6**: 5.

Esposito, M.S., D.T. Maleas, K.A. Bjornstad, and L.L. Holbrook. 1986. The REC46 gene of *Saccharomyces cerevisiae* controls mitotic chromosomal stability, recombination and sporulation: Cell-type and life cycle stage-specific expression of the *rec46-1* mutation. *Curr. Genet.* **10**: 425.

Esposito, M.S., J. Hosoda, J. Golin, H. Moise, K. Bjornstad, and D. Maleas. 1984. Recombination in *Saccharomyces cerevisiae*: REC-gene mutants and DNA-binding proteins. *Cold Spring Harbor Symp. Quant. Biol.* **49**: 41.

Hosoda, J., L.L. Holbrook, H. Moise, K. Bjornstad, D. Maleas, and M.S. Esposito. 1985. Fractionation of DNA metabolic proteins of *Saccharomyces cerevisiae* by DNA cellulose chromatography: SSB-1, SS-DNA dependent ATPase, DNA polymerase, DNA primase, topoisomerase I, and resolvase. In *The Upjohn/Labatt Symposium on the Biochemistry and Molecular Biology of Industrial Yeasts* (ed. G. Stuart and R. Klein). CRC Press, Boca-Raton, Florida. (In press.)

Symington, L.S., P.T. Morrison, and R. Kolodner. 1984. Genetic recombination catalyzed by cell-free extracts of *Saccharomyces cerevisiae*. *Cold Spring Harbor Symp. Quant. Biol.* **49**: 805.

Association of Reciprocal Exchange and Intrachromosomal Gene Conversion in Mitosis

H.L. Klein and K.K. Willis
Department of Biochemistry
New York University Medical Center, New York, New York 10016

Intrachromosomal gene conversion is a process that occurs between duplicated genes in both meiosis and mitosis. The frequency of intrachromosomal gene conversion is approximately the same as the frequency of gene conversion between allelic sequences on homologous chromosomes (Klein and Petes 1981; Jackson and Fink 1985). However, in contrast to gene conversion between homologs, intrachromosomal gene conversion has not been found to be associated with reciprocal exchange in the flanking sequences (Klein and Petes 1981; Jackson and Fink 1981, 1985; Klein 1984).

This problem has been studied further in mitosis through the use of heteroallelic duplications. Prototrophic segregants arising from two different heteroallelic duplications have been examined. The prototrophic class represents a subset of all the possible interactions between the heteroalleles. Two unusual classes of prototrophic segregants have been recovered. The first class involves a reciprocal exchange between the duplicated genes and occurs regardless of the orientation of the heteroalleles. The second class is a triplication, arising from an unequal exchange between sister chromatids. In some of these, the wild-type copy is in an unexpected location within the array. These events represent between 5% and 50% of the prototrophic segregants. They may be interpreted as arising from repair of heteroduplex and postdivision segregation.

As an alternative approach to the study of the association of reciprocal exchange and gene conversion, we have used a system that allows us to select first for exchange in an inverted repeat and then determine whether that exchange is associated with a gene conversion.

Intrachromosomal gene conversion occurs between duplicated genes at a frequency of 10^{-4} to 10^{-5} in mitosis. To facilitate the mitotic studies, duplications with heteroalleles were constructed at the *LEU2* or *HIS3* locus. At least one of the two alleles in each duplication involved the loss of a restriction enzyme site within the gene, allowing an unambiguous ordering of the duplication with respect to the centromere. It also permitted a rapid determination of the genotype of each gene following a conversion event where one gene remained mutant while the second gene was converted to wild type.

All of these studies were done with mitotic cultures using haploid strains. Orientation 1 of the duplication is shown at the top of Figure 1, and orientation 2 is shown below in Figure 1. At least 36 independent prototrophic segregants were analyzed from each duplication in both orientations. Four strains in total were examined. Three types of prototrophic segregants were recovered; those with a single copy of a wild-type gene, those with two genes, one wild type and one with one of the original mutations, and triplications.

The segregants with duplications represent simple gene conversions. They occurred in all four constructions. The order of the alleles had no effect on the gene conversion; that is, in either orientation, either gene was converted to wild type with approximately equal frequency.

Orientation 1 gave single-copy, wild-type segregants that must have arisen from reciprocal exchange in the duplications. These represent between 25% and 50% of the prototrophic segregants in a single experiment. The *HIS3* duplication also gave four prototrophic segregants that had three copies of the gene, with the wild-type copy in the middle.

Orientation 2 gave slightly different results. Prototrophic segregants that still had duplicated genes showed conversion of either allele with approximately equal frequency. No other classes were seen with the *LEU2* construction. However, the *HIS3* duplication gave both triplications and single-copy, wild-type chromosomes. In orientation 2 the triplication is expected to be the result of unequal sister chromatid exchange. The single-copy, wild-type gene is unexpected.

The percentage reciprocal exchange to give a single-copy gene is much higher than has been reported in a heteroallelic duplication at the *HIS4* locus (Jackson and Fink 1981). The single-copy segregants in orientation 1 can be explained as crossovers between the two alleles, although intragenic reciprocal exchange is rare in yeast (Roman 1957). If this were the case, then the triplications from orientation 2 should occur at an equal frequency. As is shown in Figure 1, this was not observed. An additional problem is the overall frequency of reciprocal exchange. In a similar construction at *HIS4*,

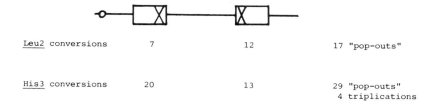

Leu2 conversions   7   12   17 "pop-outs"

His3 conversions   20   13   29 "pop-outs"
                                                           4 triplications

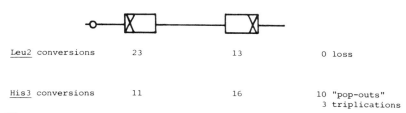

Leu2 conversions   23   13   0 loss

His3 conversions   11   16   10 "pop-outs"
                                                           3 triplications

Figure 1 Prototrophic segregants from orientations 1 and 2 of *LEU2* or *HIS3* heteroallelic duplications. (*Top*) Orientation 1. Under each mutant gene, illustrated as a box with an X in it, is listed the number of conversions of that allele to wild type with the second allele remaining mutant. The right-hand column lists the number of single gene segregants ("pop-outs") and triplications. (*Bottom*) Orientation 2 and the recovered prototrophic segregants.

the percentage of single-copy prototrophic segregants was 12% (Jackson and Fink 1981). This may be related to the distance between the heteroalleles. In the experiments reported here the *leu2* mutations are at least 400 bp apart and could be as much as 900 bp. The *his3* mutations are 320 bp apart.

The second unexpected aspect of these segregants is the high frequency of recovery of classes that are not easily explained by reciprocal exchanges that have an associated gene conversion. These are triplications from orientation 1 and the single-copy segregants from orientation 2. The recovery of these classes may be explained by the formation of heteroduplex covering at least one of the two alleles. Different constraints must be placed on the different recombination models to permit heteroduplex formation involving at least one allele. The heteroduplex must not be repaired but instead is resolved by replication. The Holliday junction is resolved in the crossover mode at a high frequency. The distance between the heteroalleles within the gene may be critical in the recovery of these unexpected crossover products. If the alleles are far apart, conversion for one allele will not involve the second allele in the heteroduplex. This is being tested using different heteroalleles with different spacings.

Direct selection for crossovers in repeated sequences is best done with inverted repeats so that all the products of the crossover can be recovered for analysis. We have recently developed a system where a gene is flanked by inverted repeats. In one orientation the gene is not efficiently expressed, but following a crossover in the inverted repeats the gene is reoriented and now is efficiently expressed. The gene that has been used for these studies is Tn*903*. This sequence encodes kanamycin resistance, which in yeast can be monitored as resistance to G418. The Tn*903* is flanked by the inverted repeats IS*903*. Tn*903* with 360 bp of IS*903* inverted repeats has been inserted into two different locations on chromosome III.

To measure gene conversion associated with a crossover, a restriction site heterology has been introduced into one of the inverted repeats. From 25% to 50% of the crossovers are associated with a gene conversion of the heterology. This means that either crossovers can occur in the absence of gene conversion or that the gene conversion tracts may be very small in these inverted repeats and so may not always involve the heterologous site. A third explanation would be restoration of the parental genotype after heteroduplex formation. All the observed gene conversions, 11 conversions in 32 crossovers, are all in one direction to the wild-type sequence.

DISCUSSION

Recent studies on duplications in mitotic growth have found that in certain constructions a high percentage of intrachromosomal events can be recovered that are associated with a crossover. Many of these are not readily predicted from a crossover between two mutant alleles or a coconversion of the two alleles with an exchange in the flanking region. Selection of crossovers in inverted repeats with subsequent assessment of whether a gene conversion has occurred in a nearby region has shown that at most half of the crossovers are associated with a conversion. These results suggest that some aspects of mitotic recombination may be different in duplicated sequences from the recombinations that occur between homologous chromosomes.

ACKNOWLEDGMENTS

This work was supported by National Institutes of Health grant GM30439. H.L.K. was a recipient of an American Cancer Society Junior Faculty Award.

REFERENCES

Jackson, J.A. and G.R. Fink. 1981. Gene conversion between duplicated genetic elements in yeast. *Nature* **292**: 306.

———. 1985. Meiotic recombination between duplicated genetic elements in *Saccharomyces cerevisiae*. *Genetics* **109**: 303.

Klein, H.L. 1984. Lack of association between intrachromosomal gene conversion and reciprocal exchange. *Nature* **310**: 748.

Klein, H.L. and T.D. Petes. 1981. Intrachromosomal gene conversion in yeast. *Nature* **289**: 144.

Roman, H. 1957. Studies of gene mutation in *Saccharomyces*. *Cold Spring Harbor Symp. Quant. Biol.* **21**: 175.

High-frequency Meiotic Recombination Events Do Not Require End-to-end Chromosome Synapsis

S. Jinks-Robertson and T.D. Petes
Department of Molecular Genetics and Cell Biology
University of Chicago, Chicago, Illinois 60637

Recombination between homologous chromosomes is a feature of meiosis in almost all eukaryotes. It has been suggested (for review, see Baker et al. 1976) that it is necessary to have at least one crossover per pair of synapsed chromosomes to ensure proper chromosome disjunction. Most researchers believe that efficient meiotic recombination requires formation of a synaptonemal complex between the interacting chromosomal DNA sequences (for review, see Von Wettstein et al. 1984). In most diploid organisms, these complexes form exclusively between homologous chromosomes, although in organisms capable of haploid meiosis, synaptonemal complex formation is sometimes observed between nonhomologous chromosomes (Von Wettstein et al. 1984).

If the synaptonemal complex is required for high levels of meiotic recombination and if synaptonemal complexes in diploid organisms do not form between regions of homology on nonhomologous chromosomes, one would predict that repeated genes on nonhomologous chromosomes would recombine infrequently relative to allelic sequences. Below, we describe evidence indicating that such recombination events in the yeast *Saccharomyces cerevisiae* are not infrequent.

RESULTS
Recombination events can be either reciprocal or nonreciprocal (gene conversions). The specific type of recombination that we investigated is meiotic recombination between repeated sequences on nonhomologous chromosomes. We will first describe studies of

meiotic gene conversion between repeated sequences and then discuss experiments designed to examine reciprocal recombination between repeated genes.

Meiotic Gene Conversion between Repeated Genes on Nonhomologous Chromosomes

We constructed two related diploid yeast strains, SJR24 and SJR33 (Jinks-Robertson and Petes 1985). The SJR24 diploid was homozygous for the *ura3-52* mutation (a Ty insertion within the *ura3* coding sequences [Rose and Winston 1984]) on chromosome V and a small deletion within the *his3* coding sequence on chromosome XV. The strain also contained two mutant *suc2* genes on chromosome IX. One mutant gene contained the wild-type *URA3* gene inserted into a restriction site in the coding sequence of *suc2* (*suc2::URA3*); the *suc2* mutant allele on the other homologous chromosome (*suc2::HIS3*) was constructed by insertion of the wild-type *HIS3* gene into the same restriction site as that used for *URA3*. Thus, the strain is heterozygous for mutant alleles of *suc2* caused by insertion of different selectable genes.

We examined tetrads derived from SJR24 (Jinks-Robertson and Petes 1985). As expected, for almost all tetrads examined, two of the spores were His⁺Ura⁻ and two of the spores were His⁻Ura⁺, indicating 2:2 segregation of the *suc2* heteroalleles. We found, however, that three tetrads (of a total of 606) segregated 1 His⁻Ura⁻ spore:1 His⁺Ura⁻ spore:2 His⁻Ura⁺ spores. By Southern analysis of DNA isolated from the His⁻Ura⁻ spore, we showed that the wild-type *HIS3* information inserted within the *suc2* locus on chromosome IX had been replaced by mutant information derived from the mutant *his3* information on chromosome XV. This transfer of information was not reciprocal since the mutant *his3* genes on chromosome XV were unchanged by this interaction. Thus, at a frequency of about 0.5%, we observed meiotic gene conversion between repeated sequences on nonhomologous chromosomes. As a control for this experiment, we constructed a strain (SJR33) which was isogenic with SJR24 except for the position of the *his3* sequences. In SJR33, the wild-type *HIS3* gene was allelic to the *his3* deletion on chromosome XV. In 690 tetrads derived from SJR33, we observed 11 conversion events. In summary, the frequency of meiotic gene conversion between repeated genes on nonhomologous chromosomes is not much less than the frequency of meiotic conversion between allelic sequences. Because of the small number of conversion tetrads, however, there could be as much as a 28-fold difference in the frequency of these two types of events.

Reciprocal Recombination between Repeated Genes on Nonhomologous Chromosomes

"Normal" allelic gene conversion events are frequently associated with reciprocal recombination of flanking markers (for review, see Fogel et al. 1981). In the study described above, we did not detect any reciprocal recombination events. Because of the small number of conversion events detected, however, this result is not necessarily significant. To investigate more definitively the association between conversion and reciprocal exchange for repeated genes on nonhomologous chromosomes, we constructed a strain (SJR59) that was homozygous for the *ura3-50* mutation on chromosome V and had one copy of a *ura3-3* mutant gene inserted within the *his3* locus on chromosome XV. The two *ura3* heteroalleles were placed in the same orientation relative to their respective centromeres. The frequency of meiotic gene conversion was estimated by the frequency of Ura$^+$ spores (5×10^{-5}). To determine whether gene conversion between the *ura3* heteroalleles was associated with reciprocal exchange, we isolated DNA from 94 Ura$^+$ spores and examined it by Southern analysis. In SJR59 (before recombination), the *ura3-3* mutant allele on chromosome XV is located on a 16-kb *Eco*RI fragment and the *ura3-50* mutant allele on chromosome V is located on a 13-kb *Eco*RI fragment. If reciprocal recombination is associated with the conversion in the Ura$^+$ spores, two different *Eco*RI fragments, 9 kb and 20 kb in size, should be produced. A total of 15 of the 94 Ura$^+$ spores (16%) contained these rearranged fragments. Genetic analysis confirms that these spores contain reciprocal translocations between chromosomes XV and V. If the reciprocal recombination events that produce the translocations occur after meiotic DNA synthesis and if the translocated chromosomes disjoin randomly, we expect that only in one-quarter of these recombination events will the spores contain the balanced translocations. Thus, we believe that about 64% ($4 \times 16\%$) of the conversion events in SJR59 are associated with reciprocal recombination of flanking DNA.

A control strain (SJR58) was constructed that was isogenic to SJR59 except that the *ura3-3* sequences were placed allelically to those of *ura3-50* on chromosome V. We found that the meiotic conversion frequency of these sequences was only 17-fold higher than observed previously for the same sequences on nonhomologous chromosomes. The association of reciprocal exchange with conversion was approximately the same for the strains SJR58 and SJR59.

We also examined the frequency of mitotic conversion and reciprocal recombination in strains SJR58 and SJR59. The mitotic frequencies for both types of recombination in both strains are lower than the meiotic frequencies by two to three orders of magnitude.

DISCUSSION

Since our results indicate that meiotic recombination events between repeated genes on nonhomologous chromosomes are very frequent, we conclude that either synaptonemal complex formation is not necessary for high levels of meiotic recombination or synaptonemal complexes occur between small regions of homology on nonhomologous chromosomes. Since no complexes between nonhomologous yeast chromosomes have been observed (Byers and Goetsch 1975), we believe the first explanation is more likely. Our results suggest that the most important factor in determining high levels of meiotic recombination in yeast is the extent of sequence homology of the interacting sequences rather than the involvement of the interacting sequences in a synaptonemal complex. One interpretation of our results is that sequence-specific pairing in meiosis is analogous to a DNA–DNA renaturation experiment. Sequences on homologous chromosomes would tend to recombine more frequently, since a recombination event at one position in the DNA would hold together the interacting chromosomes, thereby increasing the probability of a second interaction.

Because we have detected a high frequency of recombination between duplicated sequences on nonhomologous chromosomes and because the yeast genome has a large number of dispersed repeated genes (about 40 Ty elements, for example), the relative stability of the yeast genome is somewhat inexplicable. One possible explanation of the lack of chromosomal aberrations produced by recombination between dispersed Ty elements is that Ty–Ty recombination events are specifically repressed. We are currently investigating this possibility.

ACKNOWLEDGMENTS

We thank T. Nagylaki, D. Perkins, and the members of the Petes lab for helpful discussions throughout the course of this work. The research was supported by National Institutes of Health grant GM24110 (T.D.P.) and GM09628 (S.J.-R.).

REFERENCES

Baker, B.S., A.T.C. Carpenter, M.S. Esposito, R. Easton Esposito, and L. Sandler. 1976. The genetic control of meiosis. *Annu. Rev. Genet.* **10:** 53.

Byers, B. and L. Goetsch. 1975. Electron microscopic observations on the meiotic karyotype of diploid and tetraploid *Saccharomyces cerevisiae*. *Proc. Natl. Acad. Sci.* **72:** 5056.

Fogel, S., R.K. Mortimer, and K. Lusnak. 1981. Mechanisms of meiotic gene conversion, or "wanderings on a foreign strand." In *The molecular biology of the yeast Saccharomyces: Life cycle and inheritance* (ed. J.N. Strathern et al.), p. 289. Cold Spring Harbor Laboratory, Cold Spring Harbor, New York.

Jinks-Robertson, S. and T.D. Petes. 1985. High-frequency meiotic gene conversion between repeated genes on nonhomologous chromosomes in yeast. *Proc. Natl. Acad. Sci.* **82:** 3350.

Rose, M. and F. Winston. 1984. Identification of a Ty insertion within the coding sequence of the *S. cerevisiae URA3* gene. *Mol. Gen. Genet.* **193:** 557.

Von Wettstein, D., S.W. Rasmussen, and P.B. Holm. 1984. The synaptonemal complex in genetic segregation. *Annu. Rev. Genet.* **18:** 331.

Meiotic Recombination between Dispersed Homologous Sequences in *Saccharomyces cerevisiae*

M. Lichten, R.H. Borts, and J.E. Haber
Rosenstiel Center, Brandeis University
Waltham, Massachusetts 02254

Meiotic recombination in yeast is not restricted to events involving allelic sequences, but can also occur between homologous sequences present at different locations in the genome. In *Saccharomyces cerevisiae*, meiotic recombination has been observed between dispersed, repeated sequences located on the same chromosome, on homologous chromosomes, and on nonhomologous chromosomes (Petes 1980; Klein and Petes 1981; Roeder 1983; Klein 1984; Jackson and Fink 1985; Jinks-Robertson and Petes 1985). In *Schizosaccharomyces pombe*, a low level of meiotic exchange occurs between repeated tRNA genes located on nonhomologous chromosomes (Kohli et al. 1984).

In a previous examination of the timing of crossing-over in *S. cerevisiae* meiosis, we obtained physical evidence for a high level of unequal crossing-over between the normal *leu2* locus and a 2.2-kb *leu2* fragment inserted at the *MAT* locus, on the opposite arm of chromosome III (Borts et al. 1984). These observations have been extended to an examination of both allelic and unequal meiotic exchange between a pair of defined *leu2* heteroalleles inserted at various genomic locations. We have reached three conclusions: (1) The frequency of allelic exchange in an interval can be modulated by flanking sequences some distance from that interval; (2) unequal recombination between dispersed copies of a sequence can occur frequently during meiosis, often at levels comparable to that of allelic exchange; and (3) such unequal meiotic recombination events are often accompanied by crossing-over of flanking sequences.

Location Affects the Frequency of Allelic Exchange
A pair of defined mutant alleles of *leu2*, *leu2-K* and *leu2-R*, were created in vitro and introduced at the normal *leu2* locus on chromosome III. These mutant alleles were also inserted by integration

of *leu2*-containing pBR322 plasmids at three other locations: *his4* (about 20 kb centromere-distal to *leu2* on the left arm of chromosome III); *MAT* (about 100 kb from *leu2* on the opposite arm of chromosome III); and *ura3* (on chromosome V). The structures of these constructs, with the frequency of Leu2$^+$ recombinants recovered from crosses where *leu2-K* and *leu2-R* were present at allelic locations on parental homologs, are presented in Figure 1.

When both mutant alleles were present at the normal *leu2* locus, Leu2$^+$ recombinants were recovered in 1.3% of spores. Tetrad analysis indicated that greater than 90% (46/51) of Leu2$^+$ segregants were the products of gene conversion, rather than of crossing-over between the two mutant alleles. In addition, about 80% (40/51) were the products of events in which the *leu2-K* allele was converted to wild type. In the three crosses involving inserted copies of *leu2*, where the two mutant alleles were present in an identical 7.8-kb structure, the frequency of Leu2$^+$ recombinants recovered varied almost sevenfold (from 7.7×10^{-4} to 5.1×10^{-3}). These results provide a direct indication that flanking sequences that reside some distance outside a genetic interval can modulate the frequency of exchange events that occur in that interval.

Unequal Exchange between Dispersed Copies of *leu2* Occurs Frequently

To obtain an estimate of the relative frequency of interactions between dispersed copies of a sequence, a series of diploid strains were constructed in which the normal *leu2* locus was marked with either *leu2-K* or *leu2-R*, and the other *leu2* heteroallele was inserted at either *his4*, *MAT*, or *ura3*. Meiotic Leu2$^+$ segregants are the products of unequal exchange between the two dispersed copies of *leu2*, which share only 2.2 kb of homology. The structure of the diploids used, and the frequency of Leu2$^+$ random spores recovered, is presented in Table 1. In most cases, Leu2$^+$ unequal recombinants were recovered at a frequency comparable to that observed in the corresponding allelic crosses, where both *leu2-K* and *leu2-R* were inserted at the same locations and embedded in contiguous homology.

This observation is further illustrated in the cross diagramed below:

```
---leu2K----0----MATα - URA3 - leu2K-MATα--
---leu2K----0----MATa - URA3 - leu2R-MATa--
```

Leu2$^+$ recombinants can be produced either by allelic exchange between copies of *leu2-R* and *leu2-K* inserted at *MAT* or by unequal gene conversion events involving copies of *leu2-R* inserted at *MAT*

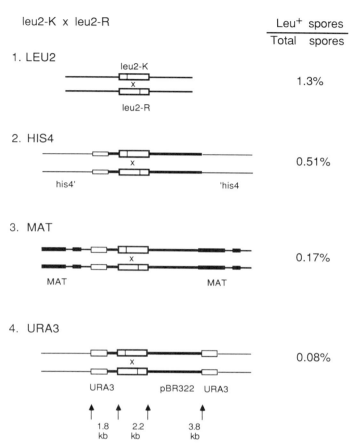

Figure 1 Allelic recombination between *leu2K* and *leu2R* inserted at various chromosomal locations. The *leu2K* and *leu2R* alleles were created by ablation of a *Kpn*I site (nucleotide 909) and an *Eco*RI site (nucleotide 1295) in the 2.2-kb *LEU2 Xho*I–*Sal*I fragment. Mutant fragments were inserted into the *Sal*I site of pBR322 plasmids which also contained the 1.2-kb *URA3* fragment at the *Hin*dIII site. Plasmid derivatives carrying a 3.5-kb *Eco*RI–*Hin*dIII *MAT* fragment or a 1.2-kb *Pvu*II–*Cla*I subfragment of the *HIS4* gene were also constructed, and were used to direct the integration of mutant copies of *leu2* to *HIS4* (cross 2), *MAT* (cross 3), and *URA3* (cross 4). In all three transformants, the *leu2* gene is embedded in 7.8 kb of identical sequence. In crosses involving inserted copies of *leu2* sequences (2, 3, and 4), the normal *LEU2* locus carried a *leu2K,R* double mutation, and therefore could not contribute to the yield of Leu2⁺ recombinants.

and *leu2-K* present at the normal chromosomal locus. Leu2⁺ spores were recovered at a frequency of 6.6×10^{-3}. About 60% of these (117/198) were conversions of *leu2-K* at the normal locus, an event that could only occur via unequal exchange. These results indicate

Table 1 Unequal Recombination between *leu2* Alleles Inserted at Different Chromosomal Locations

Diploid structure	inserted allele	frequency of Leu+ spores[a] allele at *LEU2*	
		leu2K	*leu2R*
A. HIS4			
-his4-URA3-leu2( )-his4- - - - - leu2( ) - - -•- - - - - - - - -HIS4- - - - - - - - - - - - -leu2( ) - - •- - -	*leu2K*		6.7×10^{-3} (96/237)
	leu2R	9.0×10^{-3} (119/237)	
B. MAT			
- - -leu2( )- - -•- - -MATα-URA3-leu2( )-MATα- - - - -leu2( )- - -•- - - - - - -MAT**a**- - - - - - - - - - -	*leu2K*		2.3×10^{-3}
	leu2R	9.3×10^{-3}	
C. URA3			
- - -leu2( )- - -•- - - -URA3-leu2( )-URA3-○- - - - -leu2K, R-•- - - - - - - - -ura3- - - - - - -○- -	*leu2K*		1.9×10^{-4} (9/120)
	leu2R	1.5×10^{-3} (7/120)	

One mutant *leu2* allele was present at the normal *leu2* locus; the other mutant allele was inserted in one parental chromosome at the indicated location. In the crosses involving inserts at *his4* and *MAT*, both copies of the normal *leu2* locus contained the indicated single mutation. In the crosses involving inserts at *URA3*, one parental chromosome III contained the indicated single mutation at *leu2*, while the other contained the *leu2K,R* double mutation. In many cases, most notably in the crosses in sections B and C, Leu2+ recombinants accompanied by crossing-over of flanking sequences can produce haploid-inviable chromosome rearrangements, which are not recovered in spores. Therefore, the frequencies presented are a minimum measure of the total unequal exchange events.

[a] The fraction of Leu2+ spores exhibiting unequal crossing-over of flanking sequences is shown in parentheses.

that such unequal exchange events can occur frequently, despite the fact that both interacting copies could have paired with sequences at a homologous position.

Unequal Exchange Is Associated with Crossing-over
Meiotic gene conversion events involving sequences at homologous chromosomal locations are frequently accompanied by crossing-over of flanking sequences. To determine whether unequal recombination events between dispersed copies of *leu2* were also accompanied by crossing-over, we used a combination of genetic and Southern analyses to estimate the fraction of Leu2$^+$ recombinant spores containing crossover products.

Unequal recombination between copies of *leu2* inserted at *his4* and the normal *leu2* locus is associated with crossing-over. Since the 20-kb interval between *his4* and *leu2* contains no essential sequences (Roeder 1983), events that either delete or duplicate this region are haploid viable. Thus, the majority of crossovers between these dispersed copies of *leu2* can be recovered. About half (215/474) of Leu2$^+$ recombinants were associated with crossing-over of flanking sequences (Table 1A). These unequal crossovers displayed a chromosome specificity similar to that observed by Jackson and Fink (1985) in unequal exchange events involving a tandem duplication of the *his4* locus. Unequal crossovers between dispersed copies of *leu2* on homologous, nonsister chromatids (interchromosomal crossovers) were recovered four times more frequently than were crossovers involving dispersed copies of *leu2* on either the same or on sister chromatids (intrachromosomal crossovers). In contrast, exchange events not associated with crossing-over were recovered at similar frequencies both inter- and intrachromosomally (data not shown).

Meiotic crossing-over between *leu2* sequences inserted at *MAT* and the normal *leu2* locus on the opposite arm of chromosome III occurs in about 1% of haploid genome equivalents (Borts et al. 1984). These events yield haploid inviable acentric, dicentric, and deficiency circle chromosomes, and are not recovered among Leu2$^+$ spores. We have therefore used physical analysis of DNA extracted from meiotic cells to determine the chromosome specificity of such unequal crossovers. A *Bgl*II site polymorphism present in sequences centromere distal to *leu2* [hereafter designated as *(S)leu2* or *(Y)leu2*] allows distinction between inter- and intrachromosomal crossover events in a diploid of the genotype:

```
--(Y)leu2K---0---MATα-URA3-leu2R-MATα--
--(S)leu2K---0--- ------ -MATa---- -----
```

Intra- and interchromosomal unequal crossovers between the two dispersed copies of leu2 yield a 17-kb *(Y)leu2-MATα* and a 13.5-kb *(S)leu2MATα* BglII fragment, respectively. The two products appear simultaneously and at similar levels in DNA extracted at various times during meiosis (Fig. 2), indicating that intra- and interchromosomal unequal crossovers occur at the same frequency. In this cross, nearly equal numbers of *(Y)LEU2* and *(S)LEU2* spores were recovered, indicating that both intra- and interchromosomal conversions without crossing-over occurred at similar frequencies (data not shown).

Unequal exchange between the *leu2* insert at *URA3* and the normal *leu2* locus is also associated with crossing-over (Table 1C). Crossovers between these copies of *leu2* yield a reciprocal translocation between chromosomes III and V. Both translocation products are expected to cosegregate (and thus produce a viable haploid product) in only 25% of all tetrads. Translocations were recovered in about 7% (16/240) of all Leu2$^+$ spores; therefore, we estimate that 22% of exchanges between *leu2* sequences on chromosomes III and V are associated with crossing-over.

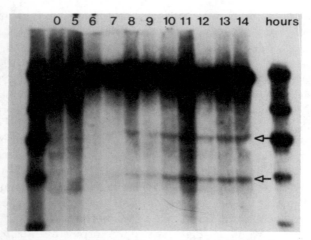

Figure 2 Detection of the products of unequal crossing-over during meiosis. DNA was extracted at the indicated times from a synchronously sporulating culture of the diploid described below, digested with BglII, displayed on a Southern blot, and probed with ^{32}P-labeled pBR322. The 17-kb *(Y)leu2*-pBR322-*MATα* and 13.5-kb *(S)leu2*-pBR322-*MATα* BglII fragments are indicated by arrows.

DISCUSSION

Meiotic recombination between dispersed copies of *leu2* sharing only 2.2 kb of homology occurs at a remarkably high level, in many cases at a frequency comparable to that of allelic exchange. A significant fraction of the observed unequal recombinants were associated with crossing-over of flanking sequences. These results indicate that extensive regions of homology are not required for efficient pairing leading to recombination, nor are they required for recombination events to be resolved as crossovers.

In *S. cerevisiae*, as in other eukaryotes, meiosis proceeds through a stage of homolog pairing and formation of synaptonemal complex (Byers and Goetch 1975; Zickler and Olson 1975). The results presented here indicate either that unequal exchange is initiated prior to this stage of meiosis, or that chromosome pairing does not preclude recombination between dispersed regions of homology.

Finally, we have observed a sevenfold variation in the frequency of allelic exchange within a defined 7.8-kb interval inserted at different chromosomal locations. Such position effects have important implications for the study of DNA sequences that stimulate or repress meiotic recombination. Our findings suggest that such elements can exert their effect over distances of at least 2000 nucleotides.

ACKNOWLEDGMENTS

This work was supported by National Institutes of Health grant GM29736 and by a grant from the March of Dimes Foundation. M.L. was supported by postdoctoral fellowships from the Damon Runyon–Walter Winchell Cancer Fund and the Leukemia Society of America. R.H.B. was supported by postdoctoral fellowships from the American Cancer Society and the Medical Foundation of Boston.

REFERENCES

Borts, R., M. Lichten, M. Hearn, M. Davidow, and J.E. Haber. 1984. Physical monitoring of meiotic recombination in *Saccharomyces cerevisiae*. *Cold Spring Harbor Symp. Quant. Biol.* **49:** 67.

Byers, B. and L. Goetch. 1975. Electron microscopic observations on the meiotic karyotype of diploid and tetraploid *Saccharomyces cerevisiae*. *Proc. Natl. Acad. Sci.* **80:** 5056.

Klein, H. 1984. Lack of association between intrachromosomal gene conversion and reciprocal exchange. *Nature* **310:** 748.

Klein, H. and T.D. Petes. 1981. Intrachromosomal gene conversion in yeast. *Nature* **289:** 144.

Kohli, J., P. Munz, R. Aebi, H. Amstutz, C. Gysler, W.-D. Heyer, L. Lehmann, P. Schuchert, P. Szankasi, P. Thuriaux, U. Leupold, J. Bell, V. Gamulin, H. Jottinger, D. Pearson, and D. Soll. 1984. Interallelic and intergenic conversion in three serine tRNA genes of *Schizosaccharomyces pombe*. *Cold Spring Harbor Symp. Quant. Biol.* **49**: 31.

Jackson, J. and G.R. Fink. 1985. Meiotic recombination between duplicated genetic elements in *Saccharomyces cerevisiae*. *Genetics* **109**: 303.

Jinks-Robertson, S. and T.D. Petes. 1985. High-frequency meiotic gene conversion between repeated genes on nonhomologous chromosomes in yeast. *Proc. Natl. Acad. Sci.* **82**: 3340.

Petes, T.D. 1980. Unequal meiotic recombination within tandem arrays of yeast ribosomal DNA genes. *Cell* **19**: 765.

Roeder, G.S. 1983. Unequal crossing-over between yeast transposable elements. *Mol. Gen. Genet.* **190**: 117.

Zickler, D. and L.W. Olson. 1975. The synaptonemal complex and the spindle plaque during meiosis in yeast. *Chromosoma* **50**: 1.

Genetic Control of Delta Recombination

J.W. Wallis, G. Chrebet, A. Beniaminovitz, and R. Rothstein
Department of Genetics and Development
College of Physicians & Surgeons
Columbia University, New York, New York 10032

Ty and delta elements comprise two families of dispersed repetitive sequences found in yeast. Ty elements are 6 kb in length, contain terminal repeats of delta sequences, and are present in 35 copies per haploid genome. Delta elements are 330 bp long and are present in 100 copies, either as part of a Ty or as solo elements (Cameron et al. 1979). Both Ty and deltas can mediate a variety of genomic rearrangements, including deletions, inversions, and translocations (for review, see Williamson 1983). We are interested in the genes that control these Ty- and delta-mediated recombination events.

To study this question we have examined deletions of the $SUP4$-o gene. We have previously shown that $SUP4$-o deletions occur at a frequency of 10^{-7} and that the end points of these deletions are within the solo delta elements that surround the locus (Rothstein 1979 and unpubl.). We describe here the isolation of a mutation, $edr1$-1, that increases the frequency of mitotic $SUP4$-o deletions by 6- to 10-fold and also increases the frequency of meiotic gene conversion at $LEU2$, a gene located near a cluster of Ty and delta elements. We also describe the cloning of $EDR1^+$ and the initial characterization of insertion mutations constructed in vitro.

RESULTS
Assay for $SUP4$-o Deletions
To assay the frequency of delta-mediated deletions at $SUP4$, it is not sufficient to look for the loss of suppressor activity since this can be due to either a deletion or a point mutation (Rothstein 1979). To distinguish $SUP4$-o deletions from the more frequently occurring point mutations, the $URA3^+$ gene was inserted next to the $SUP4$-o gene (Fig. 1). A deletion will remove both the $SUP4$-o and $URA3^+$ markers. These events are easily detected using a replica plating

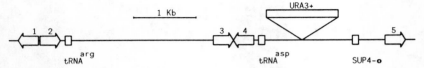

Figure 1 *SUP4*-o locus. The tyrosine-inserting tRNA suppressor gene *SUP4*-o is shown with two nearby tRNA genes, tRNA^{Arg} and tRNA^{Asp}. The position and orientation of the five solo delta elements are indicated by the large arrows. A 1.1-kb fragment containing the *URA3*+ gene was inserted at the *Hin*dIII site near *SUP4*-o.

assay. We use a strain that has the ochre-suppressible alleles *ade2-1* and *can1-100*. Suppression of these alleles causes this strain to be white and sensitive to canavanine. A deletion of *SUP4*-o causes the strain to become red, canavanine resistant, and ura3−. A point mutation eliminating suppressor activity causes the strain to become red and canavanine resistant but remain URA3+.

This assay can be done on a 3-cm² patch of cells grown on rich media. The patch is replicated to canavanine plus uracil and canavanine minus uracil media. Approximately 50 red, canavanine-resistant papillae grow up from one patch. The ura− papillae represent deletion events whereas the Ura+ point mutants serve as an internal control for the approximate number of cells in the original patch, allowing us to estimate the relative deletion frequencies.

Isolation of *edr1-1*

We mutagenized a haploid (*SUP4*-o::*URA3*+ *ade2-1 can1-100 ura3-1*) to 5% survival with ethyl methanesulfonate. The mutagenized colonies were then screened for their deletion frequency at *SUP4*. We found one mutation, *edr1-1* (enhanced delta recombination), that consistently gives a 6- to 10-fold increase in the frequency of *SUP4* deletions as measured by the patch assay. We noticed that the *edr1-1* strains grow more slowly than wild-type strains of similar genetic backgrounds.

Additionally, we found that this mutation has a meiotic recombination phenotype. The frequency of meiotic gene conversion at *leu2-3,112* is increased but conversion of other markers remains unaffected. Table 1 shows pooled data from four different crosses. In a wild-type strain the frequency of gene conversion at *leu2-3,112* is 2.9%. An *edr1-1*/+ heterozygote shows a fourfold increase in gene conversion at this locus. The control markers *leu1-12* and *trp1-1* are not affected. It is possible that the *LEU2* gene conversion events are initiated within the nearby cluster of Ty and delta elements. We

Table 1 Frequency of Meiotic Gene Conversion

	leu2-3,112	leu1-12	trp1-1
+/+	2.9% (4/136)	2.4% (1/42)	ND[a]
+/edr1-1	12.4% (18/145)	0.9% (1/111)	0 (0/48)
+/+	1.2% (1/82)	ND	ND
+/edr1-2(HIS3+)	9.8% (11/112)	ND	ND

[a]Not determined.

found that the spore viability of an edr1-1/edr1-1 homozygous diploid is too low to allow analysis of these meiotic events.

Cloning EDR1+

The approach that we took to clone $EDR1^+$ was to enrich for plasmids able to complement the growth defect of edr1-1 and then to screen these for their ability to complement the recombination phenotype. A haploid strain was transformed with a yeast library constructed in the multicopy plasmid YEp13 (the library was kindly provided by John Hill). Approximately 15,700 transformants were divided into 54 liquid cultures. The transformants in each culture were grown in log phase for 8 days while maintaining selection for the plasmids. We estimated that this growth period was sufficient for a single Edr+ transformant with a slight (20%) growth advantage to overgrow a culture. Eight single colonies from each culture were then analyzed for their SUP4-o deletion frequency as described previously, except that selection for the plasmid was maintained. In one culture we found that three out of eight transformants scored Edr+ and in another we found that one out of eight scored Edr+. Plasmids were isolated from these transformants and we determined that they all contain the identical 11.5-kb insert (Fig. 2A).

We next showed that this insert is capable of complementing the edr1-1 mitotic recombination phenotype when it is present as a single copy and that it genetically maps to the edr1-1 locus. The 9.3-kb SphI–BamHI fragment containing all but the rightmost 2.4 kb of the insert was cloned into a pUC18-$HIS3^+$ vector. Integration of the plasmid into the genome of an edr1-1 haploid was targeted by cutting within the insert at a unique NcoI site. A transformant with a single integrated copy of the plasmid has the wild-type level of SUP4-o deletions. We mated two different edr1-1 ($HIS3^+$ EDR^+) transformants to an $EDR1^+$ haploid. A total of 29 tetrads were analyzed and all showed 4:0 segregation for Edr+, demonstrating that we had cloned $EDR1^+$.

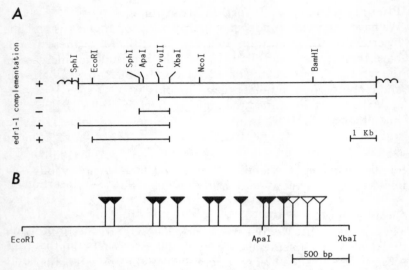

Figure 2 EDR^+ locus. (A) Restriction map of the 11.5-kb fragment that complements $edr1$-1. Wavy lines represent vector sequences. Subclones tested in the multicopy plasmid YEp13 for their ability to complement $edr1$-1 are shown as solid lines. (B) The positions of the Tn3 insertions within the 2.9-kb EcoRI–XbaI fragment are indicated. (▼) Insertions resulting in an edr1⁻ phenotype; (▽) those that result in no phenotype.

The location of the gene on the insert was determined by subcloning a series of fragments into YEp13 and also by constructing random Tn3 insertion mutations. Neither the 8.8-kb region to the right of the PvuII site nor the 0.9-kb SphI–XbaI fragment is able to complement $edr1$-1 when subcloned into YEp13. The fragments that are able to reduce the $SUP4$-o deletion frequency to wild-type levels are the leftmost 3.4 kb up to the XbaI site and the 2.9-kb EcoRI–XbaI subclone (Fig. 2A).

We used this 2.9-kb EcoRI–XbaI fragment as a target for Tn3 mutagenesis (Siefert et al. 1986), yielding 14 independent insertions. Each of the insertion mutations was analyzed in an $EDR1^+$ background by one-step gene replacement. The three insertions within 530 bp of the XbaI site did not result in an edr1⁻ phenotype. The insertion mutation 590 bp from the XbaI site resulted in a strain with high levels of $SUP4$-o deletions, as did all of the insertions to the left of this point (Fig. 2B). These insertions define a minimum size of 1.5 kb for $EDR1^+$.

Phenotypes of $edr1^-$ Insertion Mutation

We have analyzed 11 independent Tn3 insertion mutations and one insertion of $HIS3^+$ into the ApaI site (Fig. 2B). All of these strains have the identical null phenotypes, which include an increase in the frequency of delta-mediated deletions of SUP4-o, slow growth, and inability to form homozygous diploids.

We have also examined the meiotic effect of the $HIS3^+$ insertion mutation [edr1-2($HIS3^+$)] in a pair of isogenic diploids. The frequency of meiotic gene conversion at leu2-3,112 is 1.2% in an $EDR1^+/EDR1^+$ strain whereas in edr1-2($HIS3^+$)/+ this frequency is increased to 9.8% (Table 1). We have not been able to extend our analysis to a diploid homozygous for the edr1-2($HIS3^+$) insertion because we are not able to construct that strain.

Although we have not been able to construct an edr1-2($HIS3^+$)/edr1-2($HIS3^+$) diploid, we found that edr1-2($HIS3^+$) can form viable diploids with wild-type and edr1-1 strains. We also found that edr1-2($HIS3^+$) cannot form diploids with any strains containing the $edr1^-$ Tn3 insertions but that the diploids made between strains containing edr1-2($HIS3^+$) and the $Edr1^+$ insertion strains are normal. The sporulation efficiency and spore viability of a diploid heterozygous for an insertion in EDR1 is indistinguishable from the isogenic wild-type diploid.

Several other phenotypes were found for the $edr1^-$ insertion strains. The mitotic growth rate is severely limited. The doubling time is approximately three times that of a comparable wild-type strain. This difference is much more dramatic than is seen for the edr1-1 mutation. Additionally, some of the cells of an edr^- insertion strain are enlarged. The diameter of some cells is increased roughly twofold with a concomitant increase in the size of the vacuole. This defect in mitotic growth does not seem to manifest itself in other ways. The cells are still able to grow on the high osmotic media 1.5 M KCl, on the nonfermentable carbon source glycerol, and at 23°C or 36°C. They are viable after exposure to 50 Krads of γ-irradiation and an $edr1^-$ insertion mutation is viable in a $rad52^-$ background.

DISCUSSION

The effect of $edr1^-$ on the frequency of recombination events at SUP4 and at LEU2 suggest the possibility that EDR1 is specific for delta-mediated events. This is supported by preliminary experiments in which edr1-1 showed no effect on the frequency of direct repeat recombination between sequences that do not contain delta elements and also by the three- to fivefold increase in the frequency of delta–delta recombination at his4-912 seen in edr1-1 strains. EDR1 may be part of a mechanism that has evolved to suppress potentially disadvantageous recombination events between dispersed repetitive elements. Since both mitotic and meiotic recombination events are affected, we conclude that these events have some control elements in common.

An alternative to the idea that *EDR1* acts specifically on delta elements is that it suppresses recombination in specific regions of the genome. These regions, such as *SUP4*-o and *LEU2*, may accumulate delta and Ty elements because the *EDR1*-controlled depression of recombination makes these elements less likely to be involved in deleterious recombination events. This would make the presence of repeated elements in these regions more evolutionarily tolerable.

We are currently doing experiments to determine if *EDR1* functions specifically in delta- and Ty-mediated recombination or if it acts on the regions of the genome in which these elements have been found.

ACKNOWLEDGMENTS

We thank Iris Toribio for typing the manuscript. This work was supported by grants from the National Science Foundation, National Institutes of Health, and American Cancer Society.

REFERENCES

Cameron, J.R., E.Y. Loh, and R.W. Davis. 1979. Evidence for transposition of dispersed repetitive DNA families in yeast. *Cell* **16:** 739.

Rothstein, R. 1979. Deletions of a tyrosine tRNA gene in *S. cerevisiae*. *Cell* **17:** 185.

Siefert, H., E. Chen, M. So, and F. Heffron. 1986. Shuttle mutagenesis: A method of transposon mutagenesis for *Saccharomyces cerevisiae*. *Proc. Natl. Acad. Sci.* **83:** 735.

Williamson, V. 1983. Transposable elements in yeast. *Int. Rev. Cytol.* **83:** 1.

Detecting Heteroduplex DNA in Postmeiotic Segregation: And Recombination in a Nontandem *ADE8* Duplication

S. Fogel, J.W. White, A. Plessis, and D. Maloney
University of California, Department of Genetics, Berkeley, California 94720

In our attempts to develop a mechanistic description of meiotic recombination in yeast, we are concentrating on postmeiotic segregation (PMS) because we believe it is the result of a regular molecular intermediate in the recombination process. PMS is detected as half-sectoring in ascosporal colonies on only a single diagnostic medium among several used to monitor 10 heterozygous loci present in the parental diploid strain. This marker segregation at the first postmeiotic mitosis is the primary genetic evidence for heteroduplex DNA (hDNA) being delivered to a meiotic spore. The hDNA is resolved replicationally. We determined the nucleotide sequences for several heterozygous combinations that display varying PMS levels. From these sequences we developed working hypotheses describing the determinants of PMS frequency. We have tested these hypotheses by introducing systematic DNA sequence changes and analyzing their effects on PMS under conditions in which sequence context and genetic background are controlled. Finally, we have obtained the first direct physical evidence for sufficient hDNA at a particular locus in yeast ascospores to account for the observed PMS at that locus.

The DNA sequencing revealed that *ade8-18*, which displays 54% PMS among 8–10% aberrant segregations, is, in fact, a 38-bp deletion. Another high PMS marker (34% PMS), *arg4-16* is a G→C transversion. Comparison of those sequence changes (along with the *arg4-17ᵃ/arg-17ᵒ* or amber/ochre heterozygosity) and corresponding PMS frequencies with relative hDNA mismatch repair efficiencies derived by other workers on the transformation of *Streptococcus pneumoniae* suggests that PMS frequency in yeast is dictated by hDNA mismatch repair efficiency. For example, the G/T or A/C mismatch is repaired quite efficiently in *Streptococcus* and represents the lowest PMS frequency of the mismatched combinations in yeast. Also, G/G or C/C is poorly corrected in *Streptococcus* and comprises a high-frequency PMS mismatched base pair in yeast. Taken collectively, these results argue strongly that an hDNA precursor underlies most aberrant segregation events. When the hDNA contains an efficiently repaired mismatch, correction results in gene conversion (or cryptic restoration events) and PMS is infrequent. Alternately, poorly repaired hDNA mismatches or nonhomologies are delivered intact to spores more commonly to produce frequent PMS events as a consequence of resolution by DNA replication at the ensuing mitosis.

Currently, we are constructing a series of otherwise isogenic diploid yeast strains containing all possible mismatches at a single site in the *ADE8* gene. These strains contain 10 additional heterozygous markers distributed among seven chromosomes. Segregation data were collected for two such strains, i.e., for A/A or T/T mismatches, and an aberrant tetrad frequency of 14.3% was observed with 7.7% events as PMS. This finding agrees with the predicted intermediate PMS level. However, for an A/G or C/T mismatch strain, no PMS events have been observed among 41 gene conversions. Although we had no previous examples for this mismatch in yeast, analogy to *Streptococcus* predicted high PMS. This observation may reflect a difference in the enzymology of mismatch repair in these diverse organisms. Nevertheless, a wide variation in PMS frequency with a change of only a single nucleotide pair clearly implicates mismatch repair as the mediating process that determines PMS frequency.

Because *ade8-18* is a small deletion displaying very high PMS and small deletion heteroduplexes are inefficiently repaired in *Streptococcus*, we hypothesized that small deletions might constitute high PMS alleles. It was already known that frameshift mutations in several locations and large deletions in the *His4* gene do not display PMS. Therefore to investigate the relationship between deletion size and PMS frequency, we made a series of BAL-31 deletions initiated from two locations in the *ADE8* gene. Data were collected for deletions of 1, 8, 15, and 498 bp at the *Hpa*I site and 14, 38 (=*ade8-18*), 73, and 75 bp at the *Xho*I site, as well as a 1.3-kb *Kpn*I deletion. Only the 14-, 38-, and 75-bp deletions localized to the *Xho*I region display PMS, although PMS were at levels of 18, 61, and 22%, respectively. Because we see no apparent relationship between deletion size and PMS frequency, we are concentrating on a comparison of local sequences adjacent to the deletion junction points. To date, a 10-bp sequence, or close variants thereof, GGGCTGGTTA, is associated with those deletions displaying PMS. It appears that many deletions do not display PMS in yeast. Only when this 10-base sequence spans or is closely adjacent to the junction is PMS evident. We postulate that the 10-base sequence may define a binding site for a protein normally involved in other aspects of recombination, but binding fortuitously prevents hDNA correction at the deletion nonhomologies.

Physical Evidence for hDNA at the *ADE8* Locus

Physical evidence for hDNA at the *ADE8* locus in spores from an *ade8-18/ADE8+* heterozygote has been obtained by exploiting a novel restriction enzyme polymorphism. The *ade8-18* deletion removes an *Xho*I restriction site which is unique to *ADE8+* DNA while the junction sequence generates a *Bss*HII site found only in *ade8-18* DNA. Spore DNA is isolated and digested with both enzymes. Thus, homoduplex DNAs from both the homozygous mutant and wild type are cut in the center of the *ADE8* gene, but hDNA is not cleaved, since neither recognition sequence is present as double-stranded DNA. Subsequent Southern analysis reveals intense bands at 2.5 and 3.6 kb, corresponding to homoduplex. Heteroduplex is therefore expected at 2.5 + 3.6 kb = 6.1 kb. A weak band, corresponding to 1–5% of heteroduplex DNA is reproducibly observed in spore DNA from the heterozygous strain, but not from the unsporulated diploid or from spores produced by isogenic strains of *ADE8+/ADE8+* or *ade8-18/ade8-18* or mixtures thereof. This abundance of hDNA accords with the 1.3% value based on the fraction of total spores exhibiting PMS in the sporulated *ADE8/ade8-18* strain.

Recombination and Unequal Crossing-over at a Nontandem Duplication of the *ADE8* locus

Integrants of a cloned 4-kb *Ade8* fragment in Yrp17 derivatives at the *Ade8* locus were mated to produce diploids of the general genotype

(Centromere) $\dfrac{ade8\text{-}18 \quad URA3^+ \quad trp1\text{-}1 \quad ade8\text{-}18}{ade8\text{-}18 \quad ura3\text{-}1 \quad TRP1^+ \quad ADE8^+}$

Segregation of the integrated plasmids was followed in 1192 unselected tetrads with four surviving spores. DNA was isolated from the four spores of each recombinant tetrad and Southerns were probed with *Ade8* DNA.

Unequal exchanges between mispaired *Ade8* sequences occurred in ~3% of tetrads, suggesting that mispairings occur in about one-half of all meioses. Both possible mispairings occur about equally frequently. Mispairing allows creation or loss of plasmids by conversion-like events in ~1.5% of tetrads, frequently in association with conversion or PMS for the *Ade8* marker. Sister chromatid unequal exchanges and conversions occur one-tenth as often as events between homologs and are also associated with aberrant segregations of *Ade8*.

Gene conversion frequencies for *Ura3* and *Trp1* are approximately the same as at their normal centromere-linked locations. The *Ura3* fragment does not seem to stimulate recombination in adjacent intervals. Aberrant segregations for *ade8-18* are about half of their normal frequency.

ACKNOWLEDGMENTS

This research was supported by National Institutes of Health grant GM17317 and by a National Institutes of Health Biomedical Research Award.

Gene Conversion and Associated Recombination in Heterozygous Versus Heteroallelic Diploid Cells

H. Roman
University of Washington, Seattle, Washington 98195

We recently provided evidence that gene conversion could be separated temporally from crossing-over in mitotic cells (Roman and Fabre 1983). Our interpretation of the evidence was contrary to existing hypotheses designed to explain the highly correlated relationship between gene conversion and crossing-over in a nearby region (Meselson and Radding 1975; Szostak et al. 1983). Because our interpretation was regarded even by us as quite unorthodox and because our data were based on a severe selection regimen, we thought it possible that we had selected a very special sample of cells that was not representative of the majority of cells in which gene conversion could occur. Recently conducted experiments have been designed to test the temporal separation hypothesis under less stringent selection conditions.

EXPERIMENTAL PROCEDURES AND RESULTS
Platings were made on complete synthetic medium of two types of diploids. The genotype of one strain (type A) was *ADE6 CLY8 SUC1 mal1/ade6 cly8 suc1 MAL1*. Representing the dominant allele by + and the recessive by −, the genotype can be presented as + + + −/ − − − +. The other strain (type B) was *ade6-21 CLY8 SUC1 mal1/ ade6-1 cly8 suc1 MAL1*, or −+ + + −/+− − − +. One heteroallelic diploid had the *SUC1/MAL1* reversed in relation to the *CLY8* locus (as in Table 3, below). All four loci are in the right arm of chromosome VII. Also present in both diploids were *trp5* and *leu1*, near the centromere in the left arm, as shown in the heading of Table 3. These markers served as monitors of the disjunctional behavior of chromosome VII which proved in all of the data to be that expected of mitosis. In addition the diploids were homozygous for *ade2*. Therefore, the colonies of the diploids heterozygous for *ADE6* (type A) were red, whereas those from the heteroallelic diploids

(type B) were white. In the tables, only the markers in the right arm are given since these are pertinent to our inquiry.

Before plating, cells of both types were irradiated with about 3000 rads at the rate of 100 rads/second to induce red-white sectored colonies and either whole white or whole red colonies, depending on which type was irradiated. The method is nonselective except for a strong bias against symmetrical heteroduplexes at the *ADE6* locus, a bias that is minimized because, as an examination of the data shows, symmetrical heteroduplexes are relatively rare. The frequencies of sectored and other aberrant colonies are given in Table 1. We will discuss the origin of the colonies in Table 1 when we discuss Table 3. Table 2 shows what consequences can be expected if gene conversion occurred in G_1 or in G_2 as a result of asymmetric strand transfer between two duplexes of DNA.

Column 1 shows the composition of a type B diploid in G_1. Let us now assume that the third strand has undergone strand transfer so

Table 1 Frequency of Aberrant Colonies in Heterozygous and Heteroallelic Diploids

ADE6	CLY8	SUC	MAL				
+	+	+	−				
−	−	−	+				
		Frequency sectors		Whole white		Control	
C955 84-10-17		51/5146	0.99%	9/5146	0.17%	0/2738	
84-11-13		26/3343	0.78	12/3343	0.36	0/1692	
84-11-28		78/7253	1.08	16/7253	0.22	1ª/1867	
Totals		155/15742	0.98	37/15742	0.24	1/6297	0.02

ade6-21							
−1							
−	+	+	+	−			
+	−	−	−	+			
		Frequency sectors		Whole red		Control	
C965 85-4-15		19/8216	0.23	2/8216	0.024	0/2122	
C968 85-4-17		26/7764	0.33	2/7764	0.026	0/2045	
C968 85-5-2		29/8405	0.35	1/8405	0.012	1ª/1827	
−	+	+	−	+			
+	−	−	+	−			
C967 85-4-16		15/7012	0.21	0/7012	0.000	0/1849	
Totals		89/31397	0.28	5/31397	0.016	1/7843	0.013

ªSector.

Table 2 Consequences of G_1 Versus G_2 Conversion as a Result of Asymmetric Strand Transfer

G_1	G_1 conversion		G_2 chromatids		Sectored colony	Comment
		↑		↑		homoallelic
						heteroallelic
	G_2 conversion					
		↑		↑		heteroallelic

143

that the heteroallele + − replaces the − + in the second DNA strand. The void created by this transition in strand 3 is filled in by DNA synthesis copying off of strand 4. The consequences of gene conversion in G_1 are given in the two figures at the top of column 2, with the position of gene conversion denoted by a dotted line and (+). The first figure in column 2 shows conversion at the first position; the second figure shows conversion at the second position. The dotted line in the third chromatid represents the synthesis that filled the deletion produced by the gene transfer. The third column shows the completion of the events in the second column. The lines now represent chromatids rather than single DNA strands but although we are now in G_2, after DNA replication, the event that we observe occurred in G_1. The fourth column shows the type of sectored colony we would have obtained from the preceding events. It is implicit in this proposal that the single strand that undergoes strand transfer is cleaved by an endonuclease to free the two duplexes from each other. The result of conversion in G_2 between the middle pair of duplexes is shown at the bottom of the second, third, and fourth columns.

The significant point that should be stressed is that G_1 conversion produces a sectored colony in which the white side (below the horizontal line) is of two types, the top one being homoallelic and the one just below being heteroallelic; the two types are expected with equal frequencies. If, however, conversion occurs in G_2, only one type of white sector is obtained and it is the heteroallelic type. The difference between gene conversion in G_1 and G_2 was pointed out by Esposito (1978).

We now turn to what actually happens to heteroallelic diploids (type B). The results are given in Table 3. In four experiments the numbers of the genotypes (A) and (B) are given just to the right of the genotypes, that is, 10 and 3 in experiment 85-4-15. (Please note that the two symbols of the heteroallelic condition have been replaced by one in the genotype of the sectored colony so that + means *ADE6* and − can be either the − + or + − heteroallele.) The heteroallelic condition of the white sector can readily be determined by observing if red colonies are obtained after sporulation and plating of cells of the white side. If red colonies are found in the numbers expected, the white side is considered to be heteroallelic; if no reds are obtained the white side is homoallelic. It can be seen that the genotypes (A) and (B) give heteroallelic and homoallelic results with about equal frequency: 20 and 16 from genotype (A) and 8 and 14 from genotype (B). Thus, we can conclude from these results that all of the events that gave rise to genotypes (A)

Table 3 Genotypes of Red-white Sectored Colonies from Diploids

		TRP5	LEU1	ADE6 (1) I / II	CLY8 (2) III	SUC1 (3)	MAL1 (4)	
		− / +	+ / −	−+ / +−	+ / −	(+ / −)	(− / +)	Totals

		85-4-15 C965		85-4-17 C968		85-5-2 C968		85-4-16 C967		
		genotype	n	genotype	n	genotype	n	genotype	n	Totals
(A)c	Hetero a	+ + + − − − − + − + + − + − − +	10	+ + + − − − − + − + + − + − − +	7	+ + + − − − − + − + + − + − − +	13	+ + − + − − + − − + − + + − + −	6	36
	Homo b	− + + − − − − +	7	− + + − − − − +	3	− + + − − − − +	6	− + − + − − + −	4	20
		+ − − + − − − +	3	− + + − − − + +	4	− + + − − − + +	7	− + − + − − − −	2	16
			3	+ − − + − − + +	7	+ − − + − − + +	9	+ − + − − − + +	3	22
(B)	Hetero	− + + − + − − +	0	+ − − + − + + −	3	+ − − + − + + −	3	+ − + − − + − +	2	8
	Homo	− − − + − − − +	3	+ − − + + − − +	4	+ + + − + + + −	6	+ − + − + − + −	1	14
				− − − + − − − +	2	− − − + − − − +		− + − + − + − +	1	3
(C)	Hetero			− + + − − − − +	0					
	Homo			− − − + − − − +	2				1	1
(D)	Hetero					+ + + − − + + +	0		0	0
	Homo					− + + − − − + +	1		1	1
						+ + + − − − + +			1	1
(E)	Hetero					− + + − − − + −	0		0	0
	Homo					− + + − − − + −	1		1	1

Total 63
G₂ crossovers 4/63 = 6.35%

[a] Hetero, Heteroallelic ade 6-21/1.
[b] Homo, Homoallelic, either ade 6-21/-21 or 1/1.
[c] Explanation of events: (A) and (B) are convertant in G_1 for one or the other allele of ADE6; (C) conversion in G_1 as in (A) followed by a crossover in G_2 either in region I or II; (D) conversion in G_1 as in (A) followed in G_2 by a two-strand double crossover in II and III; (E) conversion in G_1 as in (A); conversion and excision-repair in G_1 at positions 3 and 4.

and (B) originated in G_1. The same is true for genotype (C) except that we have to infer in addition a crossover in G_2 between the *ADE6* locus and *CLY8*. The footnotes supply explanations for (D) and (E). The aggregate results show 4 of 63 cases that were G_2 crossovers, or 6.35%.

Now that we have established that gene conversion occurs in G_1 and crossing over in G_2, we can turn to an analysis of the results obtained in the heterozygote (type A). The data are too voluminous to be included in this brief report. They will be published elsewhere. Of the 102 cases, representing 22 different genotypes, 56 could be explained on the assumption that a crossover had occurred in G_2 between the centromere and the *ADE6* locus. Eight others require in addition one or more crossovers distal to the *ADE6* locus. Others show conversion at *ADE6*, *CLY8*, *SUC1*, and *MAL1*, either singly or multiply. These conversional events also occur in G_1. Gene conversion at *ADE6* must be $1+:3-$ by the nature of the experiment (to produce a sectored colony) but it can be in either direction at *CLY8 SUC1* or *MAL1* or the latter two can coconvert. In rare cases $4+:0-$ or $4-:0+$ segregation is obtained. Gene conversion can occur in both directions in the *same* diploid. This suggests that, while long heteroduplexes are possible, the switching of heteroduplexes from one DNA duplex to the other in the same G_1 cell may be regarded as cases of more than one strand transfer between the DNA duplexes. It is also worthy of note that of the 102 cases there were only 13 in which excision-repair of mismatches had to be assumed. It is perhaps fairer to say that there were 13 of 46 cases since the other 56 did not involve the four markers under study. The whole white colonies derived from the type-A diploid are of the genotype expected if the red side of a potentially sectored colony failed to survive at the two-cell stage.

Finally, other data show that gene conversion can indeed take place in G_2. There were 13 sectored colonies from the type-A strain that had more than one genotype in either the red side or the white side of the colony, and there were 11 such cases in the type-B strain. These gave ratios at the four loci that suggested that unrepaired heteroduplexes were common. The ratios could only have been obtained from conversional events in G_2. Thus, while gene conversion occurred predominantly in G_1, in a small number of cases it was found to occur in G_2.

DISCUSSION

Gene conversion, at least in mitosis, occurs predominantly in G_1 followed in some cases by crossing over in G_2. Mismatch repair occurs more infrequently than one might expect. Strand transfers can take place in either direction between the same two DNA duplexes. It is not unreasonable to speculate that the same events take place and are temporally separated also in meiosis. Thus, gene conversion could occur predominantly before meiotic DNA synthesis and crossing over in the four-chromatid stage. This could very well be the explanation of the results published by Sherman and Roman (1963). The recognition of homology, as indicated by gene conversion, prior to meiotic DNA replication may also play an important role in the subsequent synapsis and its consequences.

ACKNOWLEDGMENTS

I thank Michael S. Esposito for presenting this material in my absence. Mary Ruzinski was invaluable in performing the experiments. This investigation was supported by funds from the National Institutes of Health (GM 27949).

REFERENCES

Esposito, M.S. 1978. Evidence that spontaneous mitotic recombination occurs at the two-strand stage. *Proc. Natl. Acad. Sci.* **75**: 4436.

Meselson, M.S. and C.M. Radding. 1975. A general model for genetic recombination. *Proc. Natl. Acad. Sci.* **72**: 358.

Roman, H. and F. Fabre. 1983. Gene conversion and associated reciprocal recombination are separable events in vegetative cells of *Saccharomyces cerevisiae*. *Proc. Natl. Acad. Sci.* **80**: 6912.

Sherman, F. and H. Roman. 1963. Evidence for two types of allelic recombination in yeast. *Genetics* **48**: 255.

Szostak, J.W., T.L. Orr-Weaver, R.J. Rothstein, and F.W. Stahl. 1983. The double-strand-break repair model for recombination. *Cell* **33**: 25.

Synapsis-dependent Illegitimate Recombination and Rearrangement in Yeast

S. Kunes, D. Botstein, and M.S. Fox
Department of Biology, Massachusetts Institute of Technology
Cambridge, Massachusetts 02139

Mechanisms of genomic rearrangement play an important and general role in cellular DNA metabolism. Rearrangement mechanisms are implicated in modulating gene expression and in the formation of the large number of novel coding sequences employed in the immune response. Rearrangements are also known to arise in a last-ditch effort to salvage a damaged genome and by mechanisms mediated by transposable elements. A rearrangement may occur as the consequence of a normal mechanism gone awry. For example, homologous recombination between sequences on different chromosomes can result in a translocation. In some contexts, rearrangement mechanisms serve in generating the material of evolutionary change.

One particular class of DNA rearrangements is the formation of a head-to-head (inverted) duplication of DNA sequence, of which several examples have been described. The highly amplified macronuclear *Tetrahymena* rDNA is a linear inverted dimer (Engberg et al. 1976) derived from a single micronuclear germ line gene (Yao and Gall 1977). Yeast respiratory (rho⁻) mutants often harbor a highly reiterated palindromic rearrangement of a segment of the wild-type mitochondrial genome (Locker et al. 1974). The λdv plasmid derivatives of phage λ are found to often be inverted dimers of phage genomic sequence (Berg 1974; Chow et al. 1974). In maize, the daughters of a broken chromosome may fuse at their free DNA ends (McClintock 1939). The resulting dicentric chromosome can break again at mitotic anaphase, resulting in a bridge-breakage fusion cycle.

Formation of Inverted Dimer Plasmids in Yeast

We have recently reported (Kunes et al. 1984, 1985) on a novel outcome of yeast transformation that suggests the presence of a generalized rearrangement mechanism in yeast yielding a symmetrical

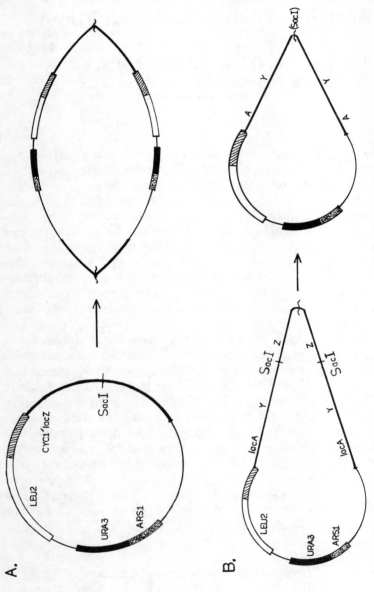

Figure 1 (See facing page for legend.)

joint at, or near, the original site of a free DNA end. A high frequency of transformants is recovered when transformation (Beggs 1978; Hinnen et al. 1978) is carried out with a mixture of sonicated nonhomologous carrier DNA and plasmid pCH308 (Fig. 1) which is linearized by SacI cleavage in its resident lacZ sequence, a sequence lacking homology with the yeast genome. Approximately 90% of these transformants harbor a head-to-head (inverted) dimer of the linearized plasmid molecules (Fig. 1). Similar results were obtained with different plasmids linearized in sequences other than lacZ, suggesting that this mode of repair is not sequence specific. The presence of the sonicated carrier DNA during transformation is apparently required for the formation of inverted dimer plasmids. In the absence of sonicated carrier DNA, transformants occur at a reduced frequency and are found to harbor a plasmid of the parental structure, or with the deletion encompassing the SacI site. The role of sonicated DNA in the formation of inverted dimer plasmids is not known.

Structure of the Novel DNA Junctions of Inverted Dimer Plasmids

When inverted dimer plasmids were examined by DNA gel-transfer analysis of yeast DNA, or by restriction analysis after isolation of the plasmids in *Escherichia coli*, deletions of typically less than several hundred basepairs were observed at the head-to-head and tail-to-tail joints. The nucleotide sequence at nine junctions was determined with plasmid DNA isolated from *E. coli* (Fig. 2). Several observations (to be described elsewhere) suggest that the structure of these junctions is that of the original junctions resulting from their primary formation in yeast. As shown in Figure 2, the junctions occur between the original sites of short (4–8 bp) homology present in inverted orientations near the SacI-generated lacZ ends. In some cases, notable partial homology between the parent sequences is

Figure 1 Formation of inverted dimer plasmids or monomer products after yeast transformation with linearized plasmid DNA. (A) Plasmid pCH308 (left of arrow; see Kunes et al. 1985, for details) and the head-to-head (inverted) dimer product (right of arrow) of transformation with SacI-linearized pCH308 DNA in the presence of sonicated carrier DNA. (B) Plasmid pSK225 (left of arrow) is a $ura3^- LEU2^+$ derivative of pCH308 in which the lacZ region of pCH308 has been replaced with a 12-kb inverted repeat of lacZYA sequence. The lacZ–lacZ junction in pSK 225 is asymmetric, allowing its propagation in *E. coli*. The monomer product of transformation with SacI-cleaved pSK225 (right of arrow) contains a novel junction typically formed within several hundred basepairs of the SacI-generated ends.

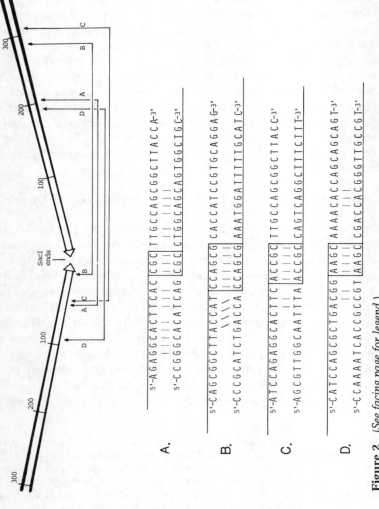

Figure 2 (See facing page for legend.)

present to either side of the site of the joint seen in the product. This junction structure resembles that of the head-to-head duplicative rearrangements of yeast mitochondrial DNA (Sor and Fukuhara 1983).

The Role of Synapsis in the Formation of Inverted Dimers

We have reported experiments indicating that, primarily, two input linearized plasmid molecules participate in the formation of each dimer plasmid (Kunes et al. 1984). When transformation is carried out at subsaturating DNA concentrations with an equimolar SacI-linearized mixture of two genetically marked derivatives of pCH308, one $ura3^-$ $LEU2^+$ and the other $URA3^+$ $leu2^-$, one-third to one-half of the resulting inverted dimer plasmids bear both the $URA3^+$ $LEU2^+$ alleles. This observation suggests that many inverted dimers are of a biparental origin. Second, the DNA concentration dependence of transformation with the linearized plasmid DNA is approximately second order, again suggesting a requirement for two molecules in the formation of a dimer plasmid. Furthermore, increasing the concentration of $URA3^+$ $leu2^-$ linear molecules in a mixture with a constant small concentration of $ura3^-$ $LEU2^+$ molecules is found to increase the yield of $Leu2^+$ transformants. This observation is consistent with the notion that $URA3^+$ $leu2^-$ linear molecules may act as partners with $ura3^-$ $LEU2^+$ linear molecules in the formation of an inverted dimer plasmid.

One hypothesis to account for the preferred head-to-head joining of two molecules is that homologous pairing brings the homologous

Figure 2 Nucleotide structure of four inverted dimer end-to-end junctions. (*Top*) The original SacI ends of two input linear molecules are shown with the molecules in the head-to-head orientation. The brackets indicate the actual sites at which these linear molecules have joined, as indicated by the nucleotide sequence analysis of junctions A, B, C, and D. The material included within each bracket is thus deleted from the product. Distances from the SacI ends are in base pairs. (*Bottom*) Nucleotide sequence of four novel junctions. The DNA sequence of both parental molecules in the region of the novel joint is shown for each junction, A, B, C, and D. The 5' to 3' parental DNA strands shown occur in an inverted orientation in lacZ near the SacI site. The nucleotide sequence of the novel joint is overlined to the left of the box of homology and underlined to the right of the box. The box of homology between the parental DNA sequences is the site of the joint. Homology between the parental DNA molecules is indicated by vertical lines at the sites of two or more contiguous matching base pairs. Junctions A and D are in the lacZ region 5' of the lacZ SacI site. Junctions B and C are in the lacZ region 3' of the lacZ SacI site.

ends into close proximity. The first evidence that homologous pairing does occur between two molecules in the formation of inverted dimers came upon examination of the heteroparental products of the transformation with a mixture of genetically marked molecules, described above. Each dimer could be expected to be heterozygous for both *URA3* and *LEU2*, with the wild-type alleles of each gene in a *trans* configuration. However, we observed that one-third of the heteroparental products were either wild-type homozygotes for *LEU2* or *URA3*, or bore the wild-type alleles in a *cis* configuration. This observation suggests that a recombinational interaction may often occur between the molecules that fuse.

If an interaction between homologous DNA sequences is the dependent step in the fusion of free DNA ends, a plasmid cleaved between an inverted repeat might be repaired by an intramolecular reaction. A direct test of this expectation was performed in the following way. The *lacZ* sequence in the genetically marked derivatives of pCH308 was replaced with an inverted repeat of *lacZYA* sequence (Fig. 1). The resulting plasmids, pSK225 and pSK227, are cleaved twice by *Sac*I, yielding molecular ends that are homologous. To test the prediction, we determined the concentration dependence of transformation with a *Sac*I-cleaved equimolar mixture of pSK225 and pSK227. A first-order DNA concentration dependence is observed with the molecules bearing intramolecular homology (data not shown), consistent with the requirement of one molecule for the formation of a transformant. Furthermore, relatively few of the transformants obtained with either *Leu2*⁺ or *Ura3*⁺ single selection were found to be both $Ura3^+$ and $Leu2^+$, unlike the outcome with the original plasmids (Table 1). Last, the $Ura3^+$ $leu2^-$ and $ura3^-$ $Leu2^+$ transformants obtained with the *Sac*I-cleaved mixture of pSK225 and pSK227 were found to harbor the monomer derivatives to be expected if the fusion reaction had been an intramolecular event (Fig. 1).

DISCUSSION

We have described the plasmid products of a novel rearrangement mechanism that yields an inverted duplication at or near the original site of a free DNA end. The nucleotide structure of the end-to-end joints of these products reveals that the end-joining process involves very short patches of homology present in an inverted orientation near the free DNA ends of the original linear plasmid molecules used in transformation.

We have shown that the majority of these dimer plasmids derive from two input linear plasmid molecules, and have provided evi-

Table 1 Transformation with a Mixture of Mutant Plasmids Linearized between a *lacZYA* Inverted Repeat: Fraction of Transformants That Are *Ura3+ Leu2+*

Primary selection	*lacZ* (**SacI*)				*lacAYZZYA* (**SacI*)			
	10[a]	25	50	100	10[a]	25	50	100
Ura3+	–	–	10/28	21/60	1/45	2/60	4/54	5/59
Leu2+	2/6	5/16	22/44	26/54	4/156	4/75	5/60	–

Yeast strain DBY1226 (*ura3-3 leu2-3,112*) was transformed by the method of Hinnen et al. (1978) at various total DNA concentrations with a *SacI*-cleaved equimolar mixture of the mutant derivatives of pCH308 (Fig. 1) or a *SacI*-cleaved equimolar mixture of the mutant plasmids harboring a *lacZYA* inverted repeat (Fig. 1). Sonicated chick erythrocyte carrier DNA (10 µg) was included in each transformation. The mutant plasmids are either *ura3⁻ Leu2+* or *Ura3+ leu2⁻*. Primary selection was carried out directly in regeneration agar after transformation. This table gives the fraction of transformants for either *Ura3+* selection or *Leu2+* selection that were found to be *Ura3+ Leu2+*. Since the linearized mixture of mutant derivatives of pCH308 gives rise to transformants with a second-order DNA concentration dependence, very few transformants are recovered with 10 ng or 25 ng of plasmid DNA.

[a] Amount of DNA mixture added (ng).

dence suggesting that homologous pairing might bring into proximity the ends that fuse. How, then, might homologous pairing be a necessary step in the joining of ends by an apparent illegitimate recombination event? One possibility is that a normal DNA intermediate in homologous recombination is a substrate for an illegitimate exchange when the recombining molecules have identical free DNA ends. A second possibility is that the illegitimate recombination that joins these molecules is a separable event from the synapsis mechanism that orients and brings into proximity the ends that fuse.

ACKNOWLEDGMENTS

This work was supported by grant AIO5388 from the National Institutes of Health (to M.S.F.). S.K. was supported by training grant GMO7287 from the National Institutes of Health.

REFERENCES

Beggs, J.D. 1978. Transformation of yeast by a replicating hybrid plasmid. *Nature* **275**: 104.

Berg, D. 1974. Genetic evidence for two types of gene rearrangements in new λdv plasmid mutants. *J. Mol. Biol.* **86**: 59.

Chow, L.T., N. Davidson, and D. Berg. 1974. Electron microscope study of the structures of λdv DNAs. *J. Mol. Biol.* **86**: 69.

Engberg, J., P. Anderson, and V. Leick. 1976. Free ribosomal DNA molecules from *Tetrahymena pyriformis* GL are giant palindromes. *J. Mol. Biol.* **104**: 455.

Hinnen, A., J.B. Hicks, and G.R. Fink. 1978. Transformation of yeast. *Proc. Natl. Acad. Sci.* **75:** 1929.

Kunes, S., D. Botstein, and M.S. Fox. 1984. Formation of inverted dimer plasmids after transformation of yeast with linearized plasmid DNA. *Cold Spring Harbor Symp. Quant. Biol.* **49:** 617.

―――――. 1985. Transformation of yeast with linearized plasmid DNA: Formation of inverted dimers and recombinant plasmid products. *J. Mol. Biol.* **184:** 375.

Locker, J., M. Rabinowitz, and G.S. Getz. 1974. Tandem inverted repeats in mitochondrial DNA of petite mutants of *Saccharomyces cerevisiae*. *Proc. Natl. Acad. Sci.* **71:** 1366.

McClintock, B. 1939. The behavior in successive nuclear divisions of a chromosome broken at meiosis. *Proc. Natl. Acad. Sci.* **25:** 405.

Sor, F. and H. Fukuhara. 1983. Unequal excision of complementary strands is involved in the generation of palindromic repetitions of rho$^-$ mitochondrial DNA in yeast. *Cell* **32:** 391.

Yao, M.-C. and J.G. Gall. 1977. A single integrated gene for ribosomal RNA in a eukaryote, *Tetrahymena pyriformis*. *Cell* **12:** 121.

Is There Distributive Pairing in *Saccharomyces cerevisiae*?

D.B. Kaback
Department of Microbiology
UMDNJ-New Jersey Medical School, Newark, New Jersey 07103

During the first meiotic division, homologous chromosomes must first pair for recombination to occur. After recombination the homologs segregate to opposite poles of the spindle apparatus. The mechanism that tells two chromosomes that they are homologous is not understood. It seems reasonable to predict that shared DNA sequences direct pairing and the processes of pairing and recombination somehow induce the subsequent segregation of the homologs. However, in *Drosophila*, two nonhomologous chromosomes that lack homologs will pair and segregate as if they were homologous, a process termed distributive pairing (Grell 1976). Since *Drosophila* chromosomes have repetitive DNA elements that are present on each of the different chromosomes (Gall et al. 1971; Wensink et al. 1973), it is possible that distributive pairing of nonhomologous chromosomes is actually due to homologous pairing of the shared repetitive elements. Alternatively, distributive pairing may be truly nonhomologous. Thus, on the basis of existing observations, it is not possible to conclude that homologous DNA governs the pairing and segregation of chromosomes during the first meiotic division.

Saccharomyces cerevisiae provides an excellent genetic system for investigating the determinants for meiotic pairing and segregation. Both circular and linear minichromosomes of various compositions can be constructed and introduced into the organism by DNA transformation (Clarke and Carbon 1980; Murray and Szostak 1983). The behavior of these minichromosomes with respect to each other or to intact chromosomes can then be examined during meiosis. By varying the composition of the minichromosomes, it should be possible to define the minimal DNA components required for proper meiotic behavior. If we assume that pairing and segregation are connected, minichromosome segregation can be monitored to determine the chromosomal elements involved in meiotic pairing. To examine the DNA requirements for meiotic pairing and segregation in *S. cerevisiae*, we investigated the meiotic behavior of several homologous and nonhomologous centromere-containing circular plasmid minichromosomes with respect to an unpaired copy of chromosome I. We found that both homologous and nonhomologous plasmids segregated from the unpaired chromosome at the first meiotic division with approximately equal efficiency. These results suggest that *S. cerevisiae* is capable of both nonhomologous distributive pairing and homologous pairing of chromosomes during meiosis.

RESULTS AND DISCUSSION

To determine if a plasmid is capable of segregating from a homologous unpaired copy of an intact chromosome at the first meiotic division, plasmid YCp50(*ADE1-CDC15*)7A was transformed into X1221a-7C, a strain monosomic for chromosome I (Bruenn and Mortimer 1970). YCp50(*ADE1-CDC15*)7A contains approximately 12 kb from the right arm of chromosome I inserted in a vector containing *CEN4*, *ARS1*, *URA3*, and pBR322 sequences (see Table 1). Transformants were sporulated and the resultant asci dissected and analyzed as described in Table 2. The plasmid appeared to segregate away from chromosome I at the first meiotic division in most of the asci examined. Of the asci, 82% fell into the 0:2 *ADE1:ade1* class, indicative of plasmid–chromosome segregation. The remaining asci showed cosegregation of the plasmid with chromosome I, giving two viable spores where either one (the 1:1 *ADE1:ade1* class) or both (the 2:0 *ADE1:ade1* class) were Ade$^+$, indicating that they contained the plasmid. In all cases the Ade$^+$ phenotype showed the appropriate mitotic instability expected for a centromere plasmid. The 1:1 class indicated cosegregation because its asci contained one live spore carrying both a copy of chromosome I and the plasmid. This class could have been caused by failure of the plasmid to replicate before meiosis, nondisjunction of the plasmid during meiosis I, or mitotic loss of the plasmid in one of the two viable spores (Clarke and Carbon 1980; Stinchcomb et al. 1982).

In this experiment, the predominance of the 0:2 class is not due to random segregation of the plasmid and chromosome. Random segregation would have made the 0:2 class equal in number to the combined 2:0 and 1:1 classes. χ^2 analysis comparing the 0:2 class with the combined other classes indicated it was extremely unlikely the plasmid segregated randomly ($p < 0.001$). Thus, a small circular plasmid appears to be capable of segregating away from an unpaired intact chromosome in a large fraction of the meiotic events analyzed.

Table 1 Plasmids Used in This Study

Plasmid	Insert description			Source
	size (kb)	chromosome	physical location	
YCp50(ADE1-CDC15)7A	12	IR	10 kb from CENI	J. Crowley and D. Kaback
YCp50(MAK16)2C	12	IL	40 kb from CENI	J. Crowley, J. ONeil, and D. Kaback
YCp50(PUT1)WB2	12	XII R	not determined[a]	S.-S. Wang and M. Brandriss
YCp50(RS)1	10	unknown		D. Kaback
YCp50(RS)2	14	unknown		D. Kaback
YCp50(λ)	16	BamHI fragment containing bacteriophage λ late genes		D. Kaback

All plasmids containing yeast DNA inserts were isolated from the library constructed by J. Thomas, M. Rose, and P. Novick. YCp50(ADE1-CDC15)7A, YCp50(MAK16)2C, and YCp50(PUT1)WB2 were isolated by complementation of appropriate S. cerevisiae mutants. YCp50(RS)1 and YCp50(RS)2 were isolated from two randomly selected colonies of E. coli strain SF8 transformed with the library. YCp50(λ) was constructed by inserting the 16-kb λ BamHI fragment into the BamHI site of YCp50.

[a] Genetically, PUT1 is located 60 cM to the right of CEN12 (S.S. Wang and M. Brandriss, pers. comm.).

Table 2 Meiotic Segregation of a Homologous *CEN* Plasmid from the Intact Copy of Chromosome I in a Strain Monosomic for That Chromosome.

Cell Type	Ascus class ADE1:ade1	Number of asci	Percentage of asci
ADE1 ade1 (diagram)	0:2	42	82%
ADE1 ade1 (diagram)	2:0	2	4%
ADE1 ade1 (diagram)	1:1	7	14%

X1221a-7c (2N-1) transformed with YCp50(*ADE1-CDC15*)7A was sporulated and analyzed genetically. The genetic consequence of each kind of segregation event is shown. The circular chromosome represents YCp50(*ADE1-CDC15*)7A, the linear chromosome is the *ade1* containing intact copy of chromosome I, happy faces indicate viable spores, and frowning faces indicate inviable spores. A total of 59 asci were dissected; 51 asci gave two viable spores that all showed first-division segregation for viability with respect to the centromere-linked *trp1* gene. Number and percentages of asci shown refer to these two viable-spore asci only. The remaining eight asci dissected gave a single viable spore that was always *ade1*. The plasmid did not significantly affect spore viability, which was approximately equal in transformed and untransformed strains.

In the experiment described above, the 0:2 ascus class attributed to plasmid–chromosome segregation could have also indicated that the plasmid was frequently lost during sporulation. We believe that this possibility is unlikely based on the stable meiotic transmission of this plasmid in related euploid strains (less than 10% of the asci showed plasmid loss). However, to show plasmid–chromosome segregation conclusively, genetic analysis was carried out using strains singly trisomic for chromosome I where all four ascospores are viable. Here, one copy of chromosome I presumably remains at least partially unpaired and segregates randomly at the first meiotic division with respect to the other two copies of chromosome I (Mortimer and Hawthorne 1973). This unpaired chromosome should be suitable for pairing with the plasmid and their meiotic segregation behavior with respect to each other can be studied. If the plasmid segregates away from this unpaired copy, asci should contain two disomic $(N+1)$ and two plasmid-bearing haploid spores $(N+P)$.

For these experiments, YCp50-based plasmids containing either homologous or nonhomologous inserts were obtained or constructed (see Table 1) and transformed into strains trisomic for chromosome I. The two homologous plasmids contained inserts from different regions of chromosome I. Nonhomologous plasmids contained either sequences from *S. cerevisiae* chromosome XII or bacteriophage λ. In addition, two randomly selected plasmids containing yeast sequences ([RS]1 and [RS]2) were picked from the YCp50 library. Based on the fraction of total genomic DNA on chromosome I, there is only a 2% chance that either of these plasmids contain sequences from this chromosome. Accordingly, it was assumed these plasmids could serve as additional nonhomologous controls. Orthogonal-field alternation gel electrophoresis (OFAGE) (Carle and Olsen 1984) experiments are in progress to show from which chromosome they are actually derived.

Transformed strains and nontransformed controls were sporulated and dissected. Segregation of the plasmid was monitored by scoring for the *URA3* gene, while haploidy and disomy were monitored as described in Table 3. In a significant majority (60–80%) of the asci that could be analyzed, both homologous and nonhomologous plasmids were found to segregate away from an intact unpaired copy of chromosome I. In these asci the plasmid was found in the haploid spores $(N+P)$, leaving two disomic spores $(N+1)$. The apparent segregation of the nonhomologous plasmids from intact chromosomes suggests an interaction between nonhomologous chromosomes in *S. cerevisiae*. As the only DNA homology between the nonhomologous plasmids and the chromosome was the related but nonidentical ∼100-bp centromere DNA sequences, we conclude that any requirement for shared homology in this kind of meiotic interaction must be this size or smaller.

The role of DNA homology and meiotic recombination in this apparent segregation cannot be assessed by these experiments. The

Table 3 First-division Meiotic Segregation of Centromere-containing Plasmids with Respect to An Extra Copy of Chromosome I

	Number of asci		
Insert	segregation away from chromosome I	cosegregation with chromosome I	Percent segregation
ADE1-CDC15(I)	116	42	73
MAK16(I)	55	28	66
PUT1(XII)	66	44	60
RS1	32	13	71
RS2	20	5	80
λ	54	30	64

Asci from trisomic strains transformed with YCp50-based plasmids containing the inserts listed below were analyzed for the presence of plasmid and the number of copies of chromosome I in each spore. If known, the chromosomal location of the plasmid insert is shown in the parentheses. Asci showing segregation of the plasmid away from chromosome I contained two disomic spores $(N + 1)$ and two spores where either one or both contained plasmid $(N$ or $N + P)$. Asci scored as showing cosegregation of the plasmid with chromosome I contained two disomic spores where either one or both contained plasmid $(N + 1 + P)$ and two haploid spores (N). First-division segregation was monitored by scoring for the centromere-linked trp1 gene. Plasmid presence was assayed by scoring the plasmid-borne URA3 gene. Only asci showing 2:2 and 1:3 segregation for URA3:ura3 were included. Virtually all the 2:2 URA3:ura3 asci showed first-division segregation. On average, 70% of the asci showed 2:2 and 6% of asci showed 1:3 segregation for URA3. The remaining 24% of the asci were either 4:0, 3:1, or 0:4 for URA3:ura3. These asci were either due to the presence of extra plasmids or plasmid loss, respectively, and could not be included in the analyses. Haploidy and disomy were assayed as described by Mortimer and Hawthorne (1973) using cdc15 and ade1 as chromosome I marker genes. Asci showing 4:0 tetrads for both of these markers could not be scored for chromosome composition and were not included in the analyses. This class comprised, on average, 20% of the asci. To score segregation of YCp50 (ADE1-CDC15)7A only, all four spores from 17 asci were crossed to haploid strains containing the cdc24 mutation (a chromosome I-linked marker gene). Each diploid was sporulated and analyzed genetically as described for the presence of an extra copy of chromosome I (Mortimer and Hawthorne 1973). Fourteen asci showed plasmid-chromosome I segregation while three asci had the plasmid and chromosome I cosegregating into the same spores. The remaining asci carrying this plasmid were monitored by measuring the disappearance of the 2:2 and 3:1; ADE1:ade1 and CDC15:cdc15 ascus classes (in prep.).

homologous plasmids segregated away from chromosome I no more often than the nonhomologous plasmids and segregation occurred in the absence of any observable recombination. However, the frequency of segregation (60–80%) was well below that found for normal chromosomes (virtually 100%). Perhaps this inefficiency is related to the lack of homologous pairing or recombination. The homologous plasmids contained only 12 kb of DNA from chromosome I. On the basis of the average relationship between physical and genetic distances (Strathern et al. 1979; H.Y. Steensma, J.C.

Crowley, and D.B. Kaback, unpubl.), this length of DNA would be expected to recombine with chromosome I only 4% of the time. This level of recombination would have a negligible effect on the percentage of asci showing segregation. In addition, asci containing plasmid-chromosome recombinants would be difficult to find. A crossover between a plasmid and a chromosome results in an unstable dicentric chromosome that would be lost rapidly (Mann and Davis 1983). Thus, in these experiments we cannot tell if recombination has any effect on the efficiency of segregation. However, based on the observed inefficiency of segregation, we can speculate that much longer stretches of homology might be required to insure that chromosomes interact homologously, recombine, and segregate efficiently.

In summary, both homologous and nonhomologous plasmids segregated with equal efficiency in the absence of recombination. Assuming pairing and segregation are linked, these results suggest there may be chromosomal interactions that do not depend on either homology or recombination. These interactions might resemble the distributive pairing of nonhomologous chromosomes observed in *Drosophila*.

ACKNOWLEDGMENTS

I am most grateful to Ed Fajardo, Vincent Guacci, and Carol Newlon for valuable discussions; Peter Novick, Joan Crowley, and Sy-Shi Wang for providing recombinant plasmids; and Jaque Lamb and Sue Coventry for technical assistance. This research was supported by a grant from the National Science Foundation.

REFERENCES

Bruenn, J. and R.K. Mortimer. 1970. Isolation of monosomics in yeast. *J. Bacteriol.* **102:** 548.

Carle, G. and M.V. Olson. 1984. Separation of chromosomal DNA molecules from yeast by orthogonal-field alternation gel electrophoresis. *Nucleic Acids Res.* **112:** 5647.

Clarke, L. and J. Carbon. 1980. Isolation of a yeast centromere and construction of functional small circular chromosomes. *Nature* **287:** 504.

Gall, J.G., E.H. Cohen, and M.L. Polan. 1971. Repetitive DNA sequences in *Drosophila*. *Chromosoma* **33:** 319.

Grell, R.F. 1976. Distributive pairing. In *The genetics and biology of* Drosophila (ed. M. Ashburner and E. Novitski), p. 435. Academic Press, New York.

Mann, C. and R.W. Davis. 1983. Instability of dicentric plasmids in yeast. *Proc. Natl. Acad. Sci.* **80:** 228.

Mortimer, R.K. and D.C. Hawthorne. 1973. Genetic mapping in *Saccharomyces* IV. Mapping of temperature-sensitive genes and use of disomic strains in localizing genes. *Genetics* **80:** 33.

Murray, A.W. and J.W. Szostak. 1983. Construction of artificial chromosomes in yeast. *Nature* **305:** 189.

Stinchcomb, D.T., C. Mann, and R.W. Davis. 1982. Centromeric DNA from *Saccharomyces cerevisiae*. *J. Mol. Biol.* **158:** 157.

Strathern, J.N., C.S. Newlon, I. Herskowitz, and J.B. Hicks. 1979. Isolation of a circular derivative of yeast chromosome III: Implications for the mechanism of mating type interconversion. *Cell* **18:** 309.

Wensink, P., D. Finnegan, J.E. Donelson, and D.S. Hogness. 1973. A system for mapping DNA sequences in the chromosomes of *Drosophila melanogaster*. *Cell* **3:** 315.

Enzymatic Systems from *Saccharomyces cerevisiae* That Catalyze the Processing of Holliday Junctions and the Repair of Mismatched Nucleotides

R. Kolodner, D. Evans, L.S. Symington, and C. Muster-Nassal
Department of Biological Chemistry, Harvard Medical School
and Laboratory of Molecular Genetics
Dana Farber Cancer Institute, Boston, Massachusetts 02115

Saccharomyces cerevisiae has been used as a model organism for studying the mechanism(s) of genetic recombination. Most of our understanding about the mechanism of genetic recombination in *S. cerevisiae* has come from genetic experiments. At present our knowledge about the enzymes that catalyze genetic recombination events and the enzymatic mechanism(s) of these recombination events is quite limited. To understand better the enzymology of genetic recombination, our laboratory has developed a cell-free system that uses extracts of mitotic *S. cerevisiae* cells to catalyze genetic recombination events between homologous plasmid DNAs and has characterized the recombination reactions catalyzed by this system in detail (Symington et al. 1983, 1984, 1985). In this communication we describe our efforts to understand how the cell-free recombination system processes DNA molecules that contain Holliday junctions and repairs DNA molecules that contain mismatched nucleotides.

Resolution of Holliday Junctions

During the analysis of the structure of the recombinants that were formed during the *S. cerevisiae* in vitro recombination reaction, we observed the formation of a DNA species that appeared to consist of two circular monomer plasmids joined together by a Holliday junction (Symington et al. 1985). Two lines of evidence suggested that the extract system contained an enzyme that resolved Holliday junctions: (1) the Holliday junction-containing DNA molecules were less stable than other recombinant species, and (2) when the Holliday junction-containing molecules were purified and reincubated in the extract system, they were converted to other recombinant species. To investigate further the existence of an enzyme or enzymes responsible for processing Holliday junctions, two model assay systems for detecting the cleavage of Holliday junctions were utilized (Fig. 1). The first assay uses the plasmid pBR322::PAL114 that contains a 114-bp palindrome (Warren and Green 1985) as substrate DNA. Such DNA molecules can be treated to extrude a 57-bp cruciform structure that contains a Holliday junction at its base. Subsequent cleavage of the Holliday junction will yield unique linear DNA molecules that can be detected using the restriction mapping assay that is illustrated in Figure 1. This type of assay has proven useful in studying the *E. coli* bacteriophage T4- and T7-encoded enzymes that cleave Holliday junctions (DeMassay et al. 1984; Kemper et al. 1984). The second type of assay illustrated in Figure 1 utilizes *E. coli* bacteriophage G4 figure-eight DNA molecules as

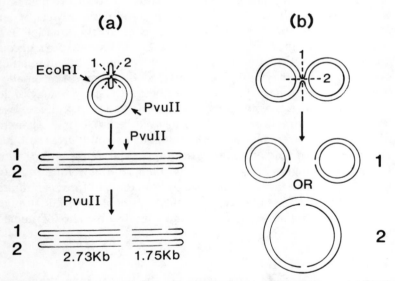

Figure 1 Illustration of assays for Holliday junction cleavage. (*a*) Partial map of pBR322::PAL114 containing an extruded cruciform structure. Cleavage of the Holliday junction in either orientation (1 or 2) yields the products labeled 1 and 2, respectively. Subsequent digestion with *Pvu*II will yield two fragments 2.73 kb and 1.75 kb long. (*b*) Structure of a figure-8 molecule. Cleavage of the Holliday junction in orientation 1 or 2 will yield circular monomers or dimers, respectively. (Reprinted, with permission, from Symington and Kolodner 1985).

substrate DNA (Thompson et al. 1976). Cleavage of the Holliday junction contained in the figure-eight molecules will yield either circular monomers or circular dimers that can be detected by either electrophoresis through agarose gels or by electron microscopy.

Using the cruciform cleavage assay, we have detected an enzymatic activity in extracts of mitotic cells that specifically makes double-strand breaks in cruciform-containing DNA molecules and have purified the activity on the order of 1000-fold (Symington and Kolodner 1985). Our best preparations of the enzyme have specific activities of approximately 350 units (cleavage of 1 pmole of cruciforms) per mg of protein and lack both endonuclease activity on double-stranded pBR322 DNA (<4 units/mg) on single-stranded M13 DNA (<5 units/mg) and exonuclease activity on double-stranded (<0.02 units/mg) and single-stranded (<0.02 units/mg) bacteriophage T7 DNA. Direct analysis of the products of the cruciform cleavage reaction by electrophoresis through acrylamide gels under denaturing conditions demonstrated that the activity cleaved the cruciforms at sites in the arms near the position of the Holliday junction and did not cleave the single-stranded regions present at the ends of the cruciform. The cruciform cleavage activity was also found to cleave the Holliday junction present in bacteriophage G4 figure-eight molecules (Thompson et al. 1976) at approximately the same rate that it cleaved a Holliday junction present at the base of a cruciform. Cleavage of the figure-eight molecules yielded primarily a mixture of circular monomers and circular dimers as well as a small proportion of α- and σ-forms. These results indicate that mitotic S. cerevisiae cells contain an enzymatic activity that cleaves Holliday junctions in the configuration that is required by a variety of recombination models to generate intact chromosomes (Holliday 1964; Meselson and Radding 1975; Szostak et al. 1983).

Correction of Mismatched Nucleotides In Vitro

Mismatched nucleotides are thought to be formed during genetic recombination, and the repair or failure to repair mismatched nucleotides during recombination has been used to explain gene conversion, postmeiotic segregation, localized negative interference, and map expansion (Holliday 1964; Meselson and Radding 1975; Fogel et al. 1979; Stahl 1979). To determine if the in vitro recombination system developed in our laboratory would repair mismatched nucleotides, we constructed a series of heteroduplex M13 substrates containing defined mismatched nucleotides and studied their repair in the in vitro recombination system (Muster-Nassal and Kolodner 1986). The mismatched nucleotides studied were 4-

bp and 7-bp insertion/deletion mismatches and each of the eight possible single-base mispairs. The substrates were designed so that the mismatched nucleotides inactivated a restriction endonuclease cleavage site which allowed us to detect mismatch repair using a restriction mapping assay, because the repair events restore the restriction endonuclease cleavage sites.

Incubation of an *Xba*I-resistant substrate DNA containing a 4-bp insertion/deletion mismatch in the in vitro recombination system for 90 minutes at 26°C resulted in the conversion of 10–20% of the substrate DNA to the *Xba*I-sensitive form in which the four inserted nucleotides had been precisely deleted. The kinetics of the repair reaction were found to be linear for 90 minutes at 26°C. The reaction had an absolute requirement for Mg^{++}, a partial requirement for rATP and the 4dNTPs, and was partially inhibited by ATPγs and completely inhibited by N-ethylmalemide.

To characterize the mismatch correction reaction further, we carried out experiments to determine the specificity of the repair reaction. The results of experiments on the specificity of mismatch repair in vitro are tabulated in Table 1. These results show that 4-bp and 7-bp insertion/deletion mismatches and AC and TG transition mispairs were repaired efficiently. The six transversion mispairs were repaired poorly or not at all. This specificity of mismatch repair in vitro is consistent with the specificity of mismatch repair in *S. cerevisiae* in vivo suggested by the data on postmeiotic segregation of markers obtained by White et al. (1985). The basis for the disparity of repair observed in some cases in vitro is not understood at present. These results indicate that the in vitro system contains enzymes that specifically recognize and repair mismatched nucleotides.

To obtain some insight into the mechanism of mismatch repair in vitro, we have characterized the mismatch repair-specific DNA synthesis that occurs during the repair reaction. Mismatch repair reactions were carried out in the presence of $\alpha^{32}P$-labeled dNTPs, and the location of regions of newly synthesized DNA within the repaired molecules was determined by restriction mapping. In the case of the efficiently repaired 4-bp and 7-bp insertion/deletion mismatches, mismatch repair-specific DNA synthesis was exclusively confined to a 54-bp region containing the mismatched nucleotides. In the case of the efficiently repaired AC mispair, mismatch repair-specific DNA synthesis was exclusively confined to a 27-bp region containing the mismatched nucleotides. In control experiments with substrates containing no mismatches or the poorly repaired TC mismatch, no mismatch-specific DNA synthesis in the vicinity of the

Table 1 Specificity of Mismatch Repair In Vitro

Substrate	Mispair	Product	Relative repair (%)	Repair events (fmole)
1	4-bp insertion/deletion	deletion	100	4.6
2	7-bp insertion/deletion	insertion	38	1.8
3	AC	AT	88	4.0
		GC	91	4.2
4	GT	GC	50	2.3
		AT	8	0.4
5	TC	TA	14	0.6
		GC	11	0.5
6	GA	GC	<3	<0.1
		TA	<3	<0.1
7	CC	CG	<3	<0.1
		GC	10	<0.5
8	GG	GC	<3	<0.1
		CG	<3	0.1
9	TT	TA	<3	0.1
		AT	20	0.9
10	AA	AT	<3	<0.1
		TA	10	0.5

Repair reactions were carried out as described by (Muster-Nassal and Kolodner 1986). An extract of AP-1 cells at a final concentration of 96 µg/ml was present in individual reactions (Symington et al. 1983, 1985). The % repair indicated is expressed relative to the frequency of repair (4.6 fmole) obtained with substrate number 1. (Reprinted, with permission, from Muster-Nassal and Kolodner 1986.)

mismatch-containing site was observed. These results indicate that in vitro, *S. cerevisiae* mismatch correction occurs by a short patch repair mechanism.

DISCUSSION

The experiments described above demonstrate that mitotic *S. cerevisiae* cells contain an enzyme that cleaves Holliday junctions in a manner that is consistent with its playing a role in genetic recombination. In addition, we have shown that mitotic cells also contain an enzymatic system that specifically recognizes and repairs mismatched nucleotides. Our results raise a number of questions that will be the subject of future investigations. In the case of the Holliday junction cleavage enzyme, we are presently purifying the enzyme to homogeneity so that we can characterize it better and study how it recognizes and interacts with Holliday junctions. We are also attempting to clone the gene that encodes the enzyme so that we can carry out genetic experiments to determine directly if the enzyme plays a role in either mitotic or meiotic recombination in *S. cerevisiae*. In the case of the mismatch repair reactions, we are now purifying the enzymes that catalyze this reaction so that we can study the enzymatic mechanism of the repair reaction. As in the case of the Holliday junction cleavage enzyme, an important priority will be to obtain genetic evidence that the mismatch repair enzymes play a role in the genetic recombination in *S. cerevisiae*.

ACKNOWLEDGMENTS

This work was supported by National Institutes of Health grant GM29383 to R.K. and by postdoctoral fellowships from the Alberta Heritage Foundation to D.E. and the Damon Runyon-Walter Winchell Cancer Fund to L.S.

REFERENCES

DeMassey, B., F.W. Studier, L. Dorgai, E. Appelbaum, and R.A. Weisberg. 1984. Enzymes and sites of genetic recombination: Studies with gene-3 endonuclease of phage T7 and site-affinity mutants of phage λ. *Cold Spring Harbor Symp. Quant. Biol.* **49:** 715.

Fogel, S., R.K. Mortimer, K. Lusnak, and V. Tavares. 1979. Meiotic gene conversion: A signal of the basic recombination event in yeast. *Cold Spring Harbor Symp. Quant. Biol.* **43:** 1325.

Holliday, R.A. 1964. A mechanism for gene conversion in fungi. *Genet. Res.* **5:** 282.

Kemper, B., F. Jensch, M.V. Depka-Prondzynski, H.-J. Fritz, V. Borgmeyer, and K. Mizuuchi. 1984. Resolution of Holliday structures by endonuclease VII as observed in interactions with cruciform DNA. *Cold Spring Harbor Symp. Quant. Biol.* **49:** 815.

Mesleson, M.S. and C.M. Radding. 1975. A general model for genetic recombination. *Proc. Natl. Acad. Sci.* **72:** 358.

Muster-Nassal, C. and R. Kolodner. 1986. Mismatch correction catalyzed by cell-free extracts of *Saccharomyces cerevisiae*. *Proc. Natl. Acad. Sci.* (in press).

Stahl, F.W. 1979. *Genetic recombination: Thinking about it in phage and fungi*. W.H. Freeman and Co., San Francisco.

Symington, L.S. and R. Kolodner. 1985. Processing of Holliday junctions by a partially purified enzymatic activity from *Saccharomyces cerevisiae*. *Proc. Natl. Acad. Sci.* **82:** 7247.

Symington, L.S., L. Fogarty, and R. Kolodner. 1983. Genetic recombination of homologous plasmids catalyzed by cell-free extracts of yeast. *Cell* **35:** 805.

Symington, L.S., P.T. Morrison, and R. Kolodner. 1984. Genetic recombination catalyzed by cell-free extracts of *Saccharomyces cerevisiae*. *Cold Spring Harbor Symp. Quant. Biol.* **49:** 805.

———. 1985. Analysis of plasmid recombination intermediates generated in a yeast cell-free recombination system. *Mol. Cell. Biol.* **5:** 2361.

Szostak, J.W., T.L. Orr-Weaver, R.J. Rothstein, and F.W. Stahl. 1983. The double-stranded break repair model of recombination. *Cell* **33:** 29.

Thompson, B.J., M.N. Camien, and R.C. Warner. 1976. Kinetics of branch migration in double-strand DNA. *Proc. Natl. Acad. Sci.* **73:** 2299.

Warren, G.J. and R.L. Green. 1985. Comparison of physical and genetic properties of palindromic DNA sequences. *J. Bacteriol.* **161:** 1103.

White, J.H., K. Lusnak, and S. Fogel. 1985. Mismatch-specific post-meiotic segregation frequency in yeast suggests a heteroduplex recombination intermediate. *Nature* **315:** 350.

Ty Element Retrotransposition

D.J. Garfinkel,* J.D. Boeke,† and G.R. Fink ‡

*NCI-Frederick Cancer Research Facility
LBI-Basic Research Program, Frederick, Maryland 21701
†Department of Molecular Biology and Genetics
Johns Hopkins School of Medicine, Baltimore, Maryland 21205
‡The Whitehead Institute for Biomedical Research
and Department of Biology
Massachusetts Institute of Technology
Cambridge, Massachusetts 02139

Ty elements comprise a class of mobile genetic elements found dispersed in the yeast genome (for review, see Roeder and Fink 1983). These 6-kb elements are similar in overall structure to certain prokaryotic transposable elements, the *copia* family of *Drosophila*, and retroviral proviruses. Terminally repeated delta (δ) sequences or long terminal repeats (LTRs) of ~335 bp surround an internal region of approximately 5.3 kb. There are about 30–50 complete elements and approximately 100 solo δs in a haploid cell. The Ty message, which is quite abundant, is transcribed from LTR to LTR forming an RNA that is terminally redundant. The recent sequencing of several Ty elements reveals two large open reading frames, *tya* and *tyb*, that take up almost all of the coding capacity of the element (Clare and Farabaugh 1985) (Fig. 1).

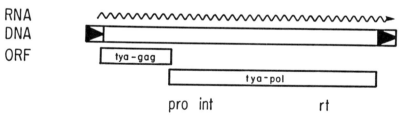

Figure 1 Structure of a Ty element. The DNA is symbolized by an open box; the triangles are LTR (δ) sequences. The wavy line symbolizes the major Ty transcript, and the two overlapping boxes represent the two open reading frames (ORFs) found in four sequenced Ty elements. *Tya* is probably analogous to the retroviral *gag* gene; the *tyb* frame is analogous to the retroviral *pol* gene. Three domains of *tyb-pol* have been identified by sequence (Mount and Rubin 1985) and functional tests. They are a protease-like domain (pro), and integrase-like domain (int), and the reverse transcriptase (polymerase) domain (rt).

Since its discovery, the study of Ty transposition was beset with many problems. A major problem is that the frequency of transposition is extremely low, around 10^{-8}/gene. Another problem in studying all repetitive elements (or genes) is that a mutation in any one element is not likely to lead to an altered phenotype because there are many other copies of the element capable of complementing the information lost by the mutant element. The endogenous elements also appear to recombine at frequencies 1000-fold greater than the transposition frequency. Homologous recombination events between disparately located Tys can lead to chromosomal aberrations such as translocations, deletions, and duplications. Furthermore, which of the many elements should be chosen for study? Some of the elements could be inactive. This certainly occurs for other elements, such as the defective *Drosophila* P elements, which require activation *in trans*. In maize, there are nonfunctioning deleted forms of *Ac* called *Ds* which only transpose in the presence of *Ac*. The challenge then is to increase the transposition frequency of Ty and then determine the mechanism by which it moves.

Revitalizing a Ty Element

Since we did not know whether a Ty element was active, we constructed a high-copy plasmid containing a fusion of a single Ty element, TyH3, to the controllable *GAL1* promoter of yeast (Boeke et al. 1985). TyH3 was chosen because it had recently transposed and it had appropriately placed restriction sites for easy plasmid construction. This construction, called pGTyH3, was genetically and physically marked with a 40-bp synthetic *lacO* sequence to distinguish where transpositions came from. Activation of the *GAL1* promoter results in as much Ty mRNA from the plasmid as is made from the entire resident population of Ty elements. On glucose this promoter is silent.

Activation of pGTyH3 dramatically increases the transposition frequency of both chromosomal Tys and the marked element. This can easily be seen using specific genes as targets for insertion, or by observing total transpositions in the entire yeast genome. One useful target gene is a promoter deletion of *His3* carried on a centromere plasmid. Scherer et al. (1982) found that at least 50% of the HIS⁺ revertants are Ty insertions that have activated *His3* transcription. In our system, essentially all of the revertants are caused by a Ty insertion with approximately 25% coming from pGTyH3.

Transpositions into yeast chromosomes also occur. We examined Ty insertions into the *LYS2* gene and found an increase of Ty-induced *lys2* mutants when TyH3 was overexpressed. Recently, W.

Thomas, J. Boeke, and F. Winston (pers. comm.) also found a high fraction of Ty insertions at *URA3* among mutants resistant to 5-fluoroorotic acid. The most pronounced demonstration that the yeast genome was suffering numerous Ty hits came from Southern filter hybridizations of total yeast DNA, isolated from cells that had survived transposition induction, with Ty or *lacO*-specific probes (Fig. 2). Each cell has numerous transpositions from the marked element and chromosomal Tys in the absence of selective pressure.

Mechanism of Transposition

The ability to induce the transposition of a marked Ty element and to follow it from the donor to the recipient site provided another key to understanding the mechanism of Ty transposition. Previous studies on other transposable elements had revealed two rather different transposition mechanisms. Bacterial transposable elements transpose DNA→DNA and the element is conserved during the process. Retroviruses transpose RNA→DNA and the element is subtly changed during the process. Reverse transcription generates DNA copies that are slightly larger than their retroviral RNA templates by virtue of the addition of segments to form the LTRs. It was possible to decide which motif was used by Ty elements by following the fate of an intron inserted into the galactose-promoted TyH3. Ty elements do not contain introns, so an intron plus its flanking exons from a yeast ribosomal protein gene were inserted into pGTyH3. When this intron-containing Ty element jumps, the intron is removed precisely by the rules of RNA splicing (Fig. 3). This result means that the Ty elements transpose DNA→RNA→DNA.

The obvious implication of an RNA intermediate in Ty transposition is that yeast cells induced for transposition contain a reverse transcriptase. Other aspects of Ty transposition support a retroviral-like reverse transcription of Ty RNA. We were fortunate in finding that the parent element contains a sequence difference between the 5' and 3' δ in the segment of the 5' δ that is present in our construction (Fig. 3). In the marked transposed Tys, the 3' δ contains the information that was present in the 5' region of the donor Ty. It appears that upon transposition the sequence from the 5' δ acts as a template for synthesis of the 3' δ. If the mRNA molecule diagrammed in Figure 3 is indeed the genetic material, that is the predicted pattern of sequence inheritance. The fact that a complete 5' δ is not required for transposition, yet is regenerated in marked transpositions, also supports the retroviral replication model.

Ty elements and retroviral proviruses are structurally related; both have LTR sequences, produce a terminally redundant LTR–

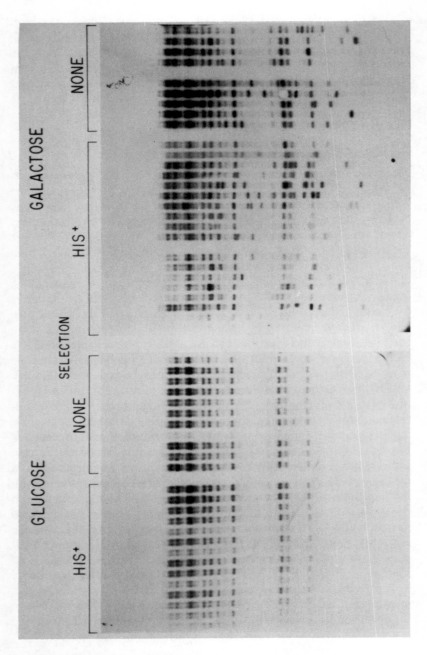

Figure 2 (*See facing page for legend.*)

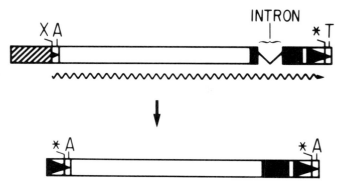

Figure 3 Features of Ty transposition events. This figure summarizes the structure of the marked *Gal1*-Ty fusions (upper segment), the corresponding RNA (wavy line), and the progeny Tys (lower segment). The hatched box is the *Gal1* promoter; the boxed arrows are δ sequences (note that about one-quarter of the 5' δ remains intact in the pGTyH3 construction); the large box is the epsilon segment; the black box symbolizes the rp51 segment. The letters A and T represent adenine and thymine residues at nucleotide 328 in the δ elements. X marks the *Xho*I site, and the asterisk shows the missing *Xho*I.

LTR transcript, have homology to a putative tRNA primer, and contain a polypurine stretch just 5' to the 3' LTR. It has recently been suggested that Ty uses a frameshifting process to express the *pol* (reverse transcriptase) gene (Clare and Farabaugh 1985) which is similar to Rous sarcoma virus. In both retroviruses and Ty, there is good evidence that the *pol* gene product is made as a readthrough product from the *gag* gene. Finally, there is significant amino acid sequence homology between the *tyb* open reading frame and retroviral and other reverse transcriptases. With all of this circumstantial evidence it was no great surprise that reverse transcriptase was found in yeast (Garfinkel et al. 1985). The unexpected result is that significant levels of reverse transcriptase are found only in yeast cells in which transposition has been induced. Uninduced yeast cells contain little or no reverse transcriptase activity. The yeast enzyme also has many of the properties of retroviral reverse transcriptase.

Figure 2 Consequences of transposition induction of Ty copy number. Southern analysis of cells grown under transposition-inducing (galactose) or noninducing conditions (glucose). Revertants of the *his3* promoter deletions were then obtained, or cells were simply exposed to galactose, recovered, and analyzed. The DNA was digested with *Hin*dIII and hybridized with a radioactively labeled *Hin*dIII–*Bgl*II fragment of TyH3.

The evolutionary relationship between Ty elements and retroviruses has become even more intimate once it was found that cells induced for Ty transposition accumulate virus-like particles (Ty-VLPs) (Garfinkel et al. 1985). The cytoplasm of transposition-induced cells is filled with spherical to ovoid particles of approximately 60 nm. The cofractionation of Ty-VLPs, reverse transcriptase activity, genome-length Ty RNA, and a *tyb*-encoded antigen and the ability of this complex to synthesize Ty DNA suggests that reverse transcription takes place in the particle. Ty-VLPs are similar morphologically and functionally to intracisternal A-type particles (IAPs) of rodents and to *copia* particles found in *Drosophila* tissue culture cells. Both IAPs and *copia* particles arise from apparent retrotransposons. The available evidence suggests that Ty elements, unlike many retroviruses, are not infectious and remain intracellular.

Mechanism of Ty Integration

Given the existing similarities between Ty elements and retroviruses, it is likely that the mechanism of Ty integration will also be similar to retroviral integration. We would then expect to find three forms of free Ty DNA in the nucleus of transposition-induced cells, genome-size Ty linears containing two LTRs, genome-size Ty circles with one LTR, and full-length circles with two tandem LTRs, which are apparently formed by blunt-end ligation of linear molecules. For one retrovirus, there is good evidence that the tandem LTR circular form is an integrative intermediate (Panganiban and Temin 1984). Placing a 50-bp restriction fragment containing an LTR–LTR junction (circle junction) from spleen necrosis virus at an internal site in the viral DNA allows integration to take place from the introduced fragment.

Panganiban (1985) has predicted the sequence of a putative Ty circle junction and compared it to the same region from other retroviruses. There is considerable sequence divergence among circle junctions, but there is one completely conserved feature. Of the dinucleotides that are eventually joined to cell DNA, 5'-TG----CA-3' are present in all retrotransposons. The enzymes responsible for Ty integration are also poorly defined. There is a weak protein homology between *tyb* and the integrase domain of the Moloney murine leukemia virus (Mo-MLV) *pol* gene (Fig. 1). No information exists concerning the roles, if any, of chromosomal genes in this process. Enzyme activities intrinsic to DNA scission, DNA synthesis, and intermolecular ligation could certainly be involved.

Effects of Ty Transposition of the Yeast Genome

Examination of a larger target, the chromosomes of transposition-induced cells, reveals numerous transposition events in virtually every cell that survives transposition induction (Fig. 2). This increase in the number of Ty elements at new locations provides many new opportunities for homologous recombination events between Ty elements. Homologous recombination between Tys has been clearly implicated in formation of genome rearrangements. Thus, a major consequence of massive transposition and enhanced homologous recombination between Ty is the destabilization of the yeast genome.

Massive retrotransposition may also have other effects that so far have not been appreciated. For example, are cellular mRNAs ever reverse-transcribed? In yeast this could create molecules that are recombinogenic. The cDNA might also be mutated because it has experienced the combined potentially mutagenic effects of RNA polymerase, reverse transcriptase (which as a group are more error prone than the DNA-dependent DNA polymerases), and exposure to a wide range of potentially damaging cellular environments than is chromosomal DNA. Yeast pseudogenes, expanded gene families, or intronless genes may be created as a consequence of massive retrotransposition. This certainly occurs in larger cells, although no such genes have yet been described in yeast. Overexpression of Ty gene products might also influence homologous recombination frequencies or DNA repair pathways.

To prevent destruction of the genome, Ty transposition normally occurs at a low frequency even though Ty mRNA is present at a high concentration. The amount of Ty mRNA produced by the induced marked element is about equal to that produced by all the chromosomal elements. For this reason, we suspect many of these chromosomal elements are defective in some way. Presumably, the overall viability of Ty elements in a cell is determined by the relative frequencies of mutation during transposition and the repair of defective elements by gene conversion. Additionally, Ty elements may have evolved a mechanism to reduce expression of Ty proteins, especially *tyb*, and they may isolate the reverse transcriptase from cellular mRNAs by sequestering it in a particle.

ACKNOWLEDGMENTS

The majority of this work was supported by grants GM35010 and CA34429 awarded to G.R.F. G.R.F. is an American Cancer Society Research Professor of Genetics. This research was sponsored in part by the National Cancer Institute, DHHS, under contract No. NO1-CO-23909 with Litton Bionetics, Inc. The contents of this publication do not necessarily reflect the views or policies of the Department of Health and Human Services, nor does mention of trade names, commercial products, or organizations imply endorsement by the U.S. Government.

REFERENCES

Boeke, J.D., D.J. Garfinkel, C.A. Styles, and G.R. Fink. 1985. Ty elements transpose through an RNA intermediate. *Cell* **40:** 491.

Clare, J. and P.J. Farabaugh. 1985. Nucleotide sequence of a Ty1 element: Evidence for a novel mechanism of gene expression. *Proc. Natl. Acad. Sci.* **82:** 2829.

Garfinkel, D.J., J.D. Boeke, and G.R. Fink. 1985. Ty element transposition: Reverse transcriptase and virus-like particles. *Cell* **42:** 507.

Mount, S.M. and G.M. Rubin. 1985. Complete nucleotide sequence of the *Drosophila* transposable element *copia*: Homology between *copia* and retroviral proteins. *Mol. Cell. Biol.* **5:** 1630.

Panganiban, A.T. 1985. Retroviral DNA integration. *Cell* **42:** 5.

Panganiban, A.T. and H.M. Temin. 1984. Circles with two tandem LTRs are precursors to integrated retrovirus DNA. *Cell* **36:** 673.

Roeder, G. and G.R. Fink. 1983. Transposable elements in yeast. In *Mobile genetic elements* (ed. J.A. Shapiro), p. 299. Academic Press, New York.

Scherer, S., C. Mann, and R.W. Davis. 1982. Reversion of a promoter deletion in yeast. *Nature* **298:** 815.

Gene Conversion Mechanisms of Punctual and Nonpunctual Mutations in *Ascobolus*

A. Nicolas, H. Hamza, and J.-L. Rossignol
Laboratoire IMG, Université Paris-Sud, 91405 Orsay Cédex, France

Genetic studies using spore color mutations in *Ascobolus* have contributed to the description of the molecular mechanism of meiotic recombination, particularly for the process of gene conversion (for review, see Rossignol et al. 1984). Extensive studies performed on the *b2* locus have demonstrated that gene conversion of punctual mutations in *Ascobolus* occurs by mismatch correction of heteroduplex DNA intermediates. In this report, we summarize our recent studies on the gene conversion behavior of nonpunctual mutations that led us to address two important questions concerning the mechanism of gene conversion in *Ascobolus:* its plurality (do one or several mechanisms lead to gene conversion) and its directionality.

Gene Conversion Patterns of Nonpunctual Heterologies

Discrete classes of gene conversion patterns were described in *Ascobolus* for induced point mutations and correlated with their chemical nature (Leblon 1972a,b). Base-substitution mutations give all types of non-Mendelian segregation (NMS) – 6:2 (6+2m and 2+6m), 5:3 (5+3m and 3+5m), and aberrant 4:4 – a type-C NMS pattern. One-base-pair addition-deletion mutations give only 6:2 asci, with mutations assumed to be single-base-pair additions preferentially converted toward mutant (type-B NMS pattern, 2+6m > 6+2m) and a single-base-pair deletion preferentially converted toward wild type (type-A NMS pattern, 6+2m > 2+6m).

In this context of an allele-specific pattern of conversion, we have studied the meiotic gene conversion pattern of seven nonpunctual heterologies located in the *b2* locus. Results are presented in Table 1. Four heterologies correspond to the crosses of *b2* nonpunctual mutations with wild type. Three heterologies were created by crossing with each other two nonpunctual mutations. The white spore G0 mutation is probably an insertion of a transposable element; G234 and G40 are two large deletions isolated as pseudo-

Table 1 Non-Mendelian Segregation Pattern of Seven Nonpunctual Heterologies in Gene $b2$

Crosses	NMSF[a] (10^{-3})	DV[b]
10 × +	39	1.2
138 × +	2	$\infty \rightarrow +$
G0 × +	76	$3.0 \rightarrow +$
G40 × +	80	1.2
10 × G234	44	1.3
G0 × G234	83	1.1
G0 × G40	70	1.1

[a]NMSF, Non-Mendelian segregation frequency per 1000 asci.
[b]DV, Disparity value: $\rightarrow +$, toward wild type; $\rightarrow m$, toward mutant.

wild-type revertants of the G0 mutation. The white spore 10 and 138 mutations are two spontaneous deletions overlapping in the middle region of the gene and extending outside of it on either side. The gene conversion patterns of the seven heterologies share two qualitative common characteristics. First, their NMS frequencies (NMSF) are lower than those of point mutations located in the same region. Second, they show no postmeiotic segregations (PMS). Besides these similarities, they exhibit a broad diversity in their conversion patterns. The decrease in NMSF is variable—only one-third with G heterologies, but fivefold with 10/+, 10/G234 and 100-fold with 138. With respect to the directionality of conversion, five of the heterologies exhibit parity (6+2m equal 2+6m) whereas two exhibit disparity in favor of the wild-type allele. G0 shows a threefold excess and there is an apparent complete bias for 138. Point mutations in this region exhibit a disparity close to 20. Our observation that G234 and G40 impose their conversional behavior on G0 but are dominated by the 10 mutation suggests the existence of complex interactions. We have studied the G234 and G40 mutations in greater detail.

G234 and G40 Deletions Impose Their Conversion Pattern to Closely Linked Punctual Mutations

Experiments showing the interaction between closely linked type-A, -B, and -C punctual mutations strongly supported the view of gene conversion of punctual mutations occurring by mismatch correction (Leblon and Rossignol 1973, 1979). We have undertaken the same experimental approach (i.e., interaction between mutations) to address the mechanism of conversion of nonpunctual heterologies. We have tested the effect of G234 and/or G40 heterologies upon the NMS patterns of tightly linked white ascospore point mutations E2 (type A), E3 (type B), and G1,24 (type C). Results presented in Table 2 show that G234/+ and G40/+ heterologies strongly modify the NMS pattern of these punctual mutations: (1) NMSF are decreased by twofold; (2) PMS are suppressed for the type-C mutations; and (3) parity is established. The NMS patterns of the punctual mutations obtained in the interactions involving G40 are those observed for G40 crossed to wild type (Table 1). These results demonstrate that G40 imposes its own gene conversion pattern. Since G234 has the same effect as that of G40, we infer that G234 has the same NMS pattern as that of G40, and that it also imposes its NMS pattern on all kinds of tightly linked punctual mutations. Thus, the action of G234 and G40 is different from that of punctual mutations: Both A- or B-type mutations and G/+ heterologies impose their conversion pattern on closely linked type-C mutations, however, G/+ heterologies also impose their own conversion pattern on closely linked type-A and -B mutations. This contrasts with type-A and -B mutations which interact mutually on their NMS pattern.

The results presented in Table 2 show that the heterozygous status of G mutations is responsible for the effects observed. When either G234 or G40 are homozygous, they have no effect upon the NMS pattern of E3 and E2. Table 2 also shows that the G234/G40 confrontation exhibits a less drastic effect upon NMSF and disparity than the single G234/+ and G40/+ heterologies. Thus, G234 and G40 are distinct mutations because they act differently from the G40/G40 and G234/G234 homozygous crosses. Taken collectively, the difference in the NMS patterns of punctual and nonpunctual mutations, and their distinct conversional behavior in interaction studies, strongly suggests that the conversion process of nonpunctual mutations such as G/+ might be distinct from those of class-A and -B punctual mutations.

Two Mechanisms for Directional Gene Conversion

To compare punctual and nonpunctual mutations further, we have studied the gene conversion directionality of G234 in asci that show evidence of heteroduplex DNA at flanking markers, using the experimental approach described by Hastings et al. (1980). This experimental scheme used two well-spaced alleles (17 and A4) which give high PMS to select tetrads with a heteroduplex covering the *b2* locus on one chromatid, and concomitantly allow one to distinguish

Table 2 Non-Mendelian Segregation Patterns of Type-A, -B, and -C Punctual Mutations in Crosses with and without G Heterologies

	Type A			Type B			Type C		
Crosses	NMSF[a] (10^{-3})	DV[b] →+	Crosses	NMSF (10^{-3})	DV →m	Crosses	NMSF (10^{-3})	DV →+	%PMS[c]
E2 × +	178	15.9	E3 × +	172	17.3	24 × +	136	4.4	38
E2 × G234	103	1.0	E3 × G234	122	1.1	24 × G234	81	1.3	1
E2 × G40	97	1.1	E3 × G40	125	1.1	24 × G40	67	1.5	1
E2 G40 × +	109	1.0	E3 G234 × +	116	1.1	G1 × +	130	1.3	91
E2 G40 × G40	173	13.3	E3 G234 × G234	152	14.0	G1 × G234	75	1.5	5
E2 G40 × G234	167	3.8	E3 G234 × G40	126	3.0	G1 × G40	71	1.1	0

[a]NMSF, Non-Mendelian segregation frequency per 1000 asci.
[b]DV, Disparity value: →+, toward wild type; →m, toward mutant.
[c]PMS, Postmeiotic segregation as percent of total NMS.

genetically the donor and the recipient strand in the heteroduplex. This scheme comprises a pair of reciprocal crosses that have the same genetic markers but differ by the relative arrangement of the middle marker under study. Such an arrangement places the mutant allele of the middle marker on the recipient strand of the heteroduplex in one cross and on the donor strand in the reciprocal cross (a versus b). Data reported in Table 3 summarize the fate of three heterologies: E1E2 (a double punctual mutation composed of two intersuppressing type-A and -B punctual mutations), the deletion G234 crossed to wild type, and G234 crossed to E1E2.

As previously reported by Hastings et al. (1980), with E1E2 (experiment I) the same directionality is observed in the two reciprocal crosses, favoring the establishment of the E1E2 genotype in the heteroduplex regardless of whether E1E2 was on the donor or the recipient strand. The directionality of gene conversion is thus "genotype directed." The behavior of the G234 heterology contrasts with that of E1E2. In experiment II, the favored genotype is different in the reciprocal crosses. G234 is in excess in cross IIa but + is in excess in cross IIb. The directionality of gene conversion of G234 is thus "donor directed." Experiment III reveals that G234 imposes its modality of conversion to E1E2. We conclude that two mechanisms for directional gene conversion exist; the donor-directed mechanism being epistatic to the genotype-directed one. We favor the idea of two mechanisms of mismatch correction.

DISCUSSION

Large heterologies in *Ascobolus* exhibit a variety of gene conversion patterns that differ from those of induced punctual mutations. We believe that these differences have two sources. One is the interference between the nonhomologous status introduced by these nonpunctual mutations and the normal formation of recombination intermediates in their vicinity. These interferences are also certainly diversified by the different location of the mutations with respect to the recombination signals operating in the vicinity and/ or the difference in size of the different mutations. In this respect, we have reported earlier the differential marker effects of punctual and nonpunctual mutations on the formation of heteroduplex and crossing over in the gene *b2* (Rossignol et al. 1984). The second source of difference is the existence of multiple pathways of gene conversion. In this report we present evidence for the existence of two mechanisms for directional gene conversion. Each pathway can lead to directionality in gene conversion. We favor the idea that

Table 3 Distribution of Donor and Recipient Genotypes for the Middle Markers (E1E2, G234) in Selected Heteroduplexes

Experiment number	Crosses		Donor strand	Recipient strand	Genotype observed (%)	
					donor	recipient
Ia	17 E1E2	A4 × + + + +	+	E1E2	32	68
Ib	17 + +	A4 × + E1E2 +	E1E2	+	77	23
IIa	17 +	A4 × + G234 +	G234	+	73	27
IIb	17 G234	A4 × + + +	+	G234	78	22
IIIa	17 E1E2	A4 × + G234 +	G234	E1E2	76	24
IIIb	17 G234	A4 × + E1E2 +	E1E2	G234	72	28

they are alternative pathways for mismatch correction of punctual and nonpunctual heterologies on heteroduplex DNA.

ACKNOWLEDGMENTS
We wish to thank Doug Treco for a critical reading of the manuscript. This investigation was made possible by support of Université Paris-Sud and CNRS (L.A. 040086).

REFERENCES
Hastings, P.J., A. Kalogeropoulos, and J.-L. Rossignol. 1980. Restoration to the parental genotype of mismatches formed in recombinant DNA heteroduplex. *Curr. Genet.* **2:** 169.

Leblon, G. 1972a. Mechanism of gene conversion in *Ascobolus immersus*. I. Existence of a correlation between the origin of mutants induced by different mutagens and their conversion spectrum. *Mol. Gen. Genet.* **115:** 36.

———. 1972b. Mechanism of gene conversion in *Ascobolus immersus*. II. The relationships between the genetic alterations in b1 or b2 mutants and their conversion spectrum. *Mol. Gen. Genet.* **116:** 322.

Leblon, G. and J.-L. Rossignol. 1973. Mechanism of gene conversion in *Ascobolus immersus*. III. The interaction of heteroalleles in the conversion process. *Mol. Gen. Genet.* **122:** 165.

———. 1979. The interaction during recombination between closely linked allelic frameshift mutant sites in *Ascobolus immersus*. II. A and B mutant sites. *Heredity* **42:** 337.

Rossignol, J.-L., A. Nicolas, H. Hamza, and T. Langin. 1984. Origins of gene conversion and reciprocal exchange in *Ascobolus*. *Cold Spring Harbor Symp. Quant. Biol.* **49:** 13.

Summary

J.N. Strathern
NCI-Frederick Cancer Research Facility
LBI-Basic Research Program, Frederick, Maryland 21701

Classical studies with *Saccharomyces cerevisiae* have made major contributions to our understanding of the patterns of genetic inheritance. Together with those from several other organisms, these observations have been incorporated into rules and models that attempt to describe the pathways by which DNA strands interact to lead to recombination and gene conversion events. Furthermore, the analysis of yeast mutants defective in recombination and DNA damage repair has been attempted to determine the roles of cellular enzymes in these processes.

Nearly 9 years have elapsed since the development of transformation in baker's yeast. About 4 years ago the tools that allow the efficient reintroduction of specific mutations into the yeast genome were developed. Combined, these techniques have allowed the isolation of genes involved in recombination and the construction of novel substrates for recombination.

No previous meeting has been devoted specifically to the advances in the understanding of recombination and gene conversion processes that have developed by the application of molecular approaches to the classically defined problems in yeast. But what did we learn? Some significant new observations were presented at this meeting, brief descriptions of which appear in this volume. This summary highlights these observations and also touches on the more general themes.

FLP-mediated recombination of 2-micron circle is described in experiments that addressed the question of what sequences are recognized by the FLP protein and what features of those sites are important (Bruckner et al.; Sadowski et al.). Recently published data from these laboratories have defined the footprint of FLP on the recombination site and the minimal sequence required for recombination (two short direct repeats with an eight-base spacer). Of particular interest in these discussions is the observation that, despite the fact that there appears to be no specific sequence requirement for the spacer (there is a size constraint), two interacting molecules must have homology in the spacer region for recombination to occur. The requirement for homology between the interacting sequences appears to precede the cleavage step! The role of such recombination in the natural biology of 2-micron circle, in vivo results addressing sequences required for flipping, and the site of the recombination associated with the flip are addressed in Volkert and Broach. Although most of the results are consistent with the idea that recombination occurs at the site of the FLP cleavages, there is evidence that some events initiated by FLP can result in products best explained by extensive gap formation (Jayaram).

Dr. Brian Sauer (pers. comm.) presented informally a site-specific recombination system with properties similar to 2-micron circle that he and his colleagues at DuPont have constructed by expressing the *cre* protein of bacteriophage P1 in yeast. They placed the *cre* gene under the control of the *GAL1* promoter and expressed *cre* in cells that had *URA3* flanked by the *lox* site at which *cre* works. Induction of *cre* resulted in efficient loss of the *URA3* gene. The expression in yeast of prokaryotic recombination genes of known function should prove a fruitful area of research.

Gene conversion mechanisms that are active in mitochondria are described in Butow et al., including comments on the omega intron of the mitochondrial 21S RNA locus. The endonuclease encoded by the omega intron cleaves 21S genes that do not have the intron, leading to gene conversion of the omega$^-$ allele to omega$^+$. This mechanism of mitotic "drive" may be a component of meiotic drive or gene conversion disparity seen in other organisms. This abstract also presents results showing that alleles of *VAR1* are not recovered at input frequency from crosses, suggesting another possible example of gene conversion disparity in mitochondria.

Several abstracts describe efforts to identify chromosomal sites involved in the initiation or stimulation of recombination. Roeder and Keil have identified a fragment of the ribosomal RNA repeats that stimulates recombination in plasmids. They have demonstrated that this fragment stimulates recombination between homologous chromosomes seven- to eightfold but only when both homologous chromosomes have the insertion. They have also defined two sequences in the fragment required for stimulation of recombination. The correlation between these sites and the sites required for transcription of the 35S precursor rRNA was one of the provocative observations presented at the meeting.

Homothallic switching is initiated by a double-stranded cut in the *MAT* (or *MAT1* in the case of *Schizosaccharomyces pombe*) locus. This cut defines the recipient of the directed gene conversion event.

These basic observations provided the background for several abstracts about both *S. cerevisiae* and *S. pombe* on control of the initiation process, genes other than the endonuclease required for switching, and the genetic and physical consequences of gene conversion and recombination events initiated by this process. Gutz et al. review analysis of the *S. pombe* mutations that affect switching (*swi*$^-$). They have divided the 10 loci into epistasis classes and have added some *rad* genes to the list of loci with defects in switching. *swi*$^-$ genes and their role in initiation and proper resolution of the switching event are discussed further in Egel. *S. pombe* differs from *S. cerevisiae* in both the distribution of switches in clonal pedigrees and the timing of the appearance of the cut in the cell cycle. Egel and Klar both discuss experiments bearing on the mechanism by which the ability to switch is conferred on a cell. The suggestion that a *cis*-acting heritable modification of the genome at *MAT1* regulates its ability to act as a substrate for the switch is another of the provocative observations described here.

The site-specific gene conversion mediated by the HO system is a key element in several areas: the genetic and physical consequences at *MAT* (Strathern et al.), the stimulation of mitotic recombination at an artificial substrate for the HO endonuclease (Nickoloff et al.), donor preference during meiosis (informal presentation, J. Haber, pers. comm.), and induction of HO-initiated recombination during meiosis (Stahl et al.). The ability to induce recombination by the action of the HO endonuclease is now well established. The more complete characterization of the pathway of such events and determination of the degree to which they mimic meiotic or mitotic recombination is still to come.

Determination of the enzymatic functions involved in recombination, a major focus of the meeting, is addressed by Wagstaff et al., Schiestl and Hastings, and Resnick et al. Wagstaff et al. have been using a system that allows the bypass of blocks to meiosis in *rad50*. In this system they monitor intrachromosomal recombination between duplicated genes. They found a delightful paradox: *rad50* has little effect on the recombination of the rRNA genes but is required for recombination between unique DNAs embedded in the rDNA sequences. The ability of these investigators to follow these processes in haploids induced for meiotic recombination facilitates the isolation and analysis of recessive mutations. Dr. Hastings presents a similar system in which he studied the recombination that occurs in *rad52*$^-$ haploids during mitosis and isolated mutants defective in this *rad52*$^-$ independent pathway.

Esposito et al. have isolated several classes of hyperrecombination and hyporecombination mutants. Of particular interest are those classes that differentially affect gene conversion or intergenic recombination. These represent another tool for exploring the enzymatic steps in these processes.

Gene conversion frequencies vary as a function of position in several loci in a fashion that suggests that these gradients reflect some important initiating or resolution step of the gene conversion process. The origin of these gradients is one of the great unanswered questions in recombination. Results with HO, omega, and FLP suggest that they can be mimicked by site-specific recombination systems. Malone et al. have focused on the cloning and characterization of the high-gene-conversion end of a gradient at the *HIS2* locus. Both parity and gene conversion frequency can be altered by insertions into this region.

Along a gene conversion gradient, the frequency of 5:3 versus 6:2 segregation varies in an allele-dependent fashion. Fogel et al. have demonstrated that one allele of *ADE8*, which has a high proportion of 5:3 segregation, is a small deletion. They present the characterization of several other small-deletion mutations of *ADE8*. Most interestingly, they used these mutations in an experiment designed to demonstrate physically that the postmeiotic segregation events reflect the formation of unrepaired heteroduplex DNA.

It has long been a paradox that recombination is efficient between homologs in yeast, but there is little recombination between dispersed reiterated repeats leading to translocations. Several laboratories are pursuing this problem. Wallis et al. describe the *edr1* mutation, which results in an elevated rate of exchange between the delta elements, and suggest that *EDR* has a role in preventing gene conversion events between delta elements (or Ty elements) from proceeding to the pathway that produces reciprocal recombinants. Roman presents evidence that gene conversion and recombination can be temporally separated. Klein and Willis, Lichten et al., and Jinks-Robertson and Petes utilize artificially created duplications to monitor intrachromosomal or interchromosomal gene conversion and associated recombination. Although some of the earlier experiments of this kind suggested that gene conversion occurred without recombination, a fairly high association of recombination was observed in these studies. (These abstracts deserve particularly careful reading so that they can be contrasted.) Some of the most spirited discussions at the meeting centered on whether these observations bore on the role of synapsis and the synaptonemal complex in promoting recombination or restricting recombination to homologs. The experiments presented in Kunes et al. suggest a different definition for synapsis. In these experiments linear DNAs cotransformed into yeast are ligated end to end to form head-to-head inverted dimers. This homology-dependent reaction is associated with additional recombination between the parental molecules.

Kaback discusses how chromosomes that do not have a homolog segregate in meiosis. The disjunction of two nonhomologous unpaired chromosomes is called distributive pairing. These experiments addressed whether the efficiency of this disjunction can be affected by added homology or recombination.

The future and the origins of this area of research are represented in three abstracts: In vitro recombination reactions, including the characterizations of extracts from mutants defective in gene conversions or recombination (Kolodner et al.); the role of retrotransposition in genome rearrangements and the enzymatic activities involved in this process (Garfinkel et al.); and experiments in *Ascobolous* designed to determine the nature of allele-specific disparity (Nicolas et al.).

Thus, these abstracts gather together investigations focused on finding out how DNA strands interact, what enzymes are involved in these reactions, and how these interactions become manifested as genetic events. Perhaps most importantly, trends in thinking and experimental approach are exposed that characterize the study of recombination in yeast at this time. This collection of extended abstracts is a time capsule that should alert the reader to stories in the making and concepts in the process of becoming conclusions.